Ordered Regression Models

Parallel, Partial, and Non-Parallel Alternatives

Chapman & Hall/CRC
Statistics in the Social and Behavioral Sciences Series

Series Editors

Jeff Gill
Washington University, USA

Steven Heeringa
University of Michigan, USA

Wim J. van der Linden
Pacific Metrics, USA

J. Scott Long
Indiana University, USA

Tom Snijders
Oxford University, UK
University of Groningen, NL

Aims and scope

Large and complex datasets are becoming prevalent in the social and behavioral sciences and statistical methods are crucial for the analysis and interpretation of such data. This series aims to capture new developments in statistical methodology with particular relevance to applications in the social and behavioral sciences. It seeks to promote appropriate use of statistical, econometric and psychometric methods in these applied sciences by publishing a broad range of reference works, textbooks and handbooks.

The scope of the series is wide, including applications of statistical methodology in sociology, psychology, economics, education, marketing research, political science, criminology, public policy, demography, survey methodology and official statistics. The titles included in the series are designed to appeal to applied statisticians, as well as students, researchers and practitioners from the above disciplines. The inclusion of real examples and case studies is therefore essential.

Published Titles

Analyzing Spatial Models of Choice and Judgment with R
David A. Armstrong II, Ryan Bakker, Royce Carroll, Christopher Hare, Keith T. Poole, and Howard Rosenthal

Analysis of Multivariate Social Science Data, Second Edition
David J. Bartholomew, Fiona Steele, Irini Moustaki, and Jane I. Galbraith

Latent Markov Models for Longitudinal Data
Francesco Bartolucci, Alessio Farcomeni, and Fulvia Pennoni

Statistical Test Theory for the Behavioral Sciences
Dato N. M. de Gruijter and Leo J. Th. van der Kamp

Multivariable Modeling and Multivariate Analysis for the Behavioral Sciences
Brian S. Everitt

Multilevel Modeling Using R
W. Holmes Finch, Jocelyn E. Bolin, and Ken Kelley

Ordered Regression Models: Parallel, Partial, and Non-Parallel Alternatives
Andrew S. Fullerton and Jun Xu

Bayesian Methods: A Social and Behavioral Sciences Approach, Third Edition
Jeff Gill

Multiple Correspondence Analysis and Related Methods
Michael Greenacre and Jorg Blasius

Applied Survey Data Analysis
Steven G. Heeringa, Brady T. West, and Patricia A. Berglund

Informative Hypotheses: Theory and Practice for Behavioral and Social Scientists
Herbert Hoijtink

Generalized Structured Component Analysis: A Component-Based Approach to Structural Equation Modeling
Heungsun Hwang and Yoshio Takane

Statistical Studies of Income, Poverty and Inequality in Europe: Computing and Graphics in R Using EU-SILC
Nicholas T. Longford

Foundations of Factor Analysis, Second Edition
Stanley A. Mulaik

Linear Causal Modeling with Structural Equations
Stanley A. Mulaik

Age–Period–Cohort Models: Approaches and Analyses with Aggregate Data
Robert M. O'Brien

Handbook of International Large-Scale Assessment: Background, Technical Issues, and Methods of Data Analysis
Leslie Rutkowski, Matthias von Davier, and David Rutkowski

Generalized Linear Models for Categorical and Continuous Limited Dependent Variables
Michael Smithson and Edgar C. Merkle

Incomplete Categorical Data Design: Non-Randomized Response Techniques for Sensitive Questions in Surveys
Guo-Liang Tian and Man-Lai Tang

Handbook of Item Response Theory, Volume 1: Models
Wim J. van der Linden

Handbook of Item Response Theory, Volume 2: Statistical Tools
Wim J. van der Linden

Handbook of Item Response Theory, Volume 3: Applications
Wim J. van der Linden

Computerized Multistage Testing: Theory and Applications
Duanli Yan, Alina A. von Davier, and Charles Lewis

Chapman & Hall/CRC
Statistics in the Social and Behavioral Sciences Series

Ordered Regression Models

Parallel, Partial, and Non-Parallel Alternatives

Andrew S. Fullerton
Oklahoma State University
Stillwater, Oklahoma

Jun Xu
Ball State University
Muncie, Indiana

CRC Press
Taylor & Francis Group
Boca Raton   London   New York

CRC Press is an imprint of the
Taylor & Francis Group, an **informa** business

A CHAPMAN & HALL BOOK

CRC Press
Taylor & Francis Group
6000 Broken Sound Parkway NW, Suite 300
Boca Raton, FL 33487-2742

© 2016 by Taylor & Francis Group, LLC
CRC Press is an imprint of Taylor & Francis Group, an Informa business

No claim to original U.S. Government works

Printed on acid-free paper
Version Date: 20160204

International Standard Book Number-13: 978-1-4665-6973-7 (Hardback)

Library of Congress Cataloging-in-Publication Data

Names: Fullerton, Andrew S. | Xu, Jun (Professor of Sociology)
Title: Ordered regression models : parallel, partial, and non-parallel
alternatives / Andrew S. Fullerton and Jun Xu.
Description: Boca Raton : Taylor & Francis, 2016. | Series: Chapman &
Hall/CRC statistics in the social and behavioral sciences | "A CRC title."
| Includes bibliographical references and index.
Identifiers: LCCN 2015045981 | ISBN 9781466569737 (alk. paper)
Subjects: LCSH: Regression analysis.
Classification: LCC QA278.2 .F85 2016 | DDC 519.5/36--dc23
LC record available at http://lccn.loc.gov/2015045981

Visit the Taylor & Francis Web site at
http://www.taylorandfrancis.com

and the CRC Press Web site at
http://www.crcpress.com

To my mother, Ede Steele, and the memory of my father, John Fullerton.

Andrew S. Fullerton

To my mother, Yaojuan Xu, and my father, Changgui Mu.

Jun Xu

Contents

Preface

This book is about regression models for ordinal outcomes, which are variables that have ordered categories but unknown spacing between categories. Ordinal outcomes are very common in the social and behavioral sciences. Self-rated health, educational attainment, income measured in intervals of $25,000, and attitudes measured on a Likert scale are just a few basic examples of ordinal variables that one may analyze using one or more of the ordered regression models we present. Our focus in this book is narrower than most books on the analysis of ordinal variables because we only consider regression-based approaches. However, we provide arguably the most comprehensive coverage of the three major classes of ordered regression models (cumulative, stage, and adjacent) and variations based on the application of the parallel regression assumption.

We assume that readers have a basic understanding of binary regression models, such as logit and probit, and of the linear regression model. Although it is relatively easy to estimate a wide variety of ordered regression models given recent advances in statistical software programs, some researchers continue to analyze ordinal outcomes using the linear or binary regression model on the grounds that results from an ordered regression model are too difficult to interpret or present effectively. In this book, we show that ordered regression models are extensions of the more familiar binary regression model and that both models rely on the same methods of estimation and interpretation. Therefore, researchers that are familiar with the binary regression model should find it relatively straightforward to apply that knowledge to the estimation and interpretation of results from ordered regression models. We illustrate the different ordered regression models using empirical examples in order to highlight several ways to interpret and present the results.

We also show that ordered regression models have clear advantages over linear and binary regression models for the analysis of ordinal outcomes. The linear regression model requires the assumption of equal spacing among categories, which is not realistic for ordinal outcomes that by definition have discrete categories with unknown spacing among them. The binary regression model requires collapsing outcome categories in order to construct a binary outcome, which results in a loss of information. Both approaches can produce very misleading results. Ordered regression models avoid these problems because they assume that the outcome is ordinal rather than continuous, and they do not require one to recode the original ordinal variable into a single binary variable.

Ordered regression models that retain the parallel regression assumption also have two advantages over the multinomial logit model, which is designed for nominal outcomes but may also be used for ordinal outcomes. Ordered regression models with the parallel assumption retain the ordinal nature of the relationships between the dependent and independent variables, and they also require fewer parameters. Multinomial logit, however, ignores the ordering of response categories and requires additional parameters, which makes it less "efficient."

The parallel regression assumption is the key assumption in traditional ordered regression models, and we devote an entire chapter of the book to this assumption (Chapter 5). We distinguish between ordered regression models that apply the parallel regression assumption to all, some, or none of the independent variables, and we refer to these as parallel, partial, and nonparallel models, respectively. There are also several classes or groups of ordered regression models based on different probabilities of interest, including cumulative, stage, and adjacent models. The so-called "ordered logit" and "ordered probit" models are actually just the same type of ordered regression model with different link functions: the cumulative model. The fact that the cumulative logit model is simply known as the "ordinal logit" model is evidence of its popularity, but there are other ordinal logit models, such as continuation ratio logit and adjacent category logit, that one should consider. Building on our earlier work (Fullerton 2009; Fullerton and Xu 2012), we provide a conceptual framework for understanding ordered regression models based on the probability of interest and on the application of the parallel regression assumption. The large number of modeling alternatives is useful because it allows one to select the most appropriate model given the type of ordinal outcome and how restrictive the parallel assumption is for each variable.

We provide brief examples in the appendices to Chapters 2, 3, and 4 that show how to estimate the models in these chapters using R and Stata. More detailed examples are available at the following Web site: http://sociology.okstate.edu/faculty-staff-directory/faculty-directory/dr-andrew-fullerton. The Web site contains JAGS, R, and Stata codes to estimate the models we present in this book. It also contains detailed examples, syntax to reproduce the results in the book, and Stata maximum likelihood estimation (MLE) routines that we created to estimate the models in Chapters 2 through 4.

We benefitted from the comments and encouragement we received from many colleagues during the process of writing this book. We thank Jeffrey Dixon, Andrew Gelman, Michael Lacy, Michael Long, Scott Long, Ann O'Connell, and Richard Williams for their comments on chapters. We appreciate their very helpful suggestions. We are indebted to Thomas Yee for providing help with our R codes using his VGAM package, John Kruschke for sharing his expertise in Bayesian statistics and for providing help with his BEST package, and Jason Doll for many informal but informative discussions and e-mail exchanges about JAGS and rjags. We also appreciate Riley Dunlap's support for this project and our earlier work on this topic. We thank Simon Cheng and David Weakliem for their feedback on our earlier work on ordered regression models, which laid the groundwork for this book. At Chapman & Hall/CRC Press, we thank our editor, John Kimmel, for his advice and support throughout this process. We also thank the anonymous reviewers for their valuable critiques. Finally, we want to thank the students from our categorical data analysis courses at Ball State University and Oklahoma State University over the last few years for their very thoughtful comments and suggestions.

Authors

Andrew S. Fullerton is an associate professor of sociology at Oklahoma State University, Stillwater, Oklahoma. His primary research interests include work and occupations, social stratification, and quantitative methods. His work has been published in journals such as *Social Forces, Social Problems, Sociological Methods & Research, Public Opinion Quarterly*, and *Social Science Research*.

Jun Xu is an associate professor of sociology at Ball State University, Muncie, Indiana. His primary research interests include Asia and Asian America, social epidemiology, and statistical modeling and programing. His work has been published in journals such as *Social Forces, Social Science & Medicine, Sociological Methods & Research, Social Science Research*, and *The Stata Journal*.

1

Introduction

1.1 Ordinal Variables versus Ordinal Models

Social and behavioral scientists often analyze ordinal outcomes, which are variables with categories ranked from low to high on a single dimension or characteristic. However, the distances between categories are unknown. For example, the General Social Survey (GSS) includes a measure of attitudes toward same-sex marriage based on the following question: "Do you agree or disagree ... homosexual couples should have the right to marry one another." The response options are based on a Likert scale: 1 = strongly disagree, 2 = disagree, 3 = neither agree nor disagree, 4 = agree, and 5 = strongly agree. The categories are ranked from low to high according to the level of support for same-sex marriage, but it is unclear whether the distances between adjacent categories are equal. Survey researchers typically assign the values 1–5 for a Likert scale question with five categories, but the difference between categories 1 and 2 is not necessarily the same as the difference between categories 2 and 3 and so forth. Additionally, ordinal variables may have multiple dimensions. In this example, the categories reflect agreement/disagreement but also the intensity of opinion: strongly versus somewhat.

The focus of this book is on regression models for ordinal outcomes. Ordinal outcomes fall within the class of categorical variables, which also includes binary and nominal variables. Binary variables have two categories, coded as 0 or 1, and nominal variables have three or more unordered categories. Variables that have ordered categories and known distances between adjacent categories are known as interval-ratio or continuous variables.*

The use of methods for continuous outcomes, such as the linear regression model, requires the assumption of meaningful category values and known distances between categories. Linear regression with an ordinal dependent variable can produce misleading results. For example, the assumption of "linearity in the parameters," or constant effects throughout the outcome distribution, may lead to biased results (Winship and Mare 1984, p. 522; Agresti 2010, p. 5). In addition, linear regression models with ordinal outcomes may have nonsensical predictions outside the observed outcome range and a nonconstant error variance. Therefore, the traditional linear regression model is not appropriate for ordinal outcomes (McKelvey and Zavoina 1975; Winship and Mare 1984; Long 1997).

The definition of an ordinal variable is relatively straightforward. However, the definition of an ordinal model is more complicated. Using a strict interpretation (e.g., Long 2015, p. 174), a model is ordinal if it constrains the effects of variables to maintain

---

* Count variables, such as the number of promotions a worker has received, also have ordered categories with known distances between them. However, count variables are considered discrete rather than continuous because there are a finite number of values. For a discussion of the connections between count models and ordinal models, see Clogg and Shihadeh (1994, p. 150).

a "stochastic ordering" (Agresti 2010, pp. 24–25), which means that the cumulative distribution function (CDF) for one group of respondents is consistently below the CDF for another group (McCullagh 1980, p. 115).* For example, if education increases support for health care spending on a Likert scale from 1 (strong opposition) to 5 (strong support) and there is a strict stochastic ordering, then respondents with 16 years of schooling will have lower cumulative probabilities for each outcome category than respondents with only 15 years of schooling. Respondents with higher levels of education will be more likely to cluster at the upper end of the scale near strong support, which means that at any given point in the distribution they will have lower cumulative probabilities (i.e., Pr $[y \leq m]$) compared to respondents with less education. The key aspect of this definition is that stochastic ordering is the result of model constraints. We will discuss one such constraint, the parallel regression assumption, briefly at the end this chapter and then in more detail in Chapter 5.

Using the strict definition, the "parallel" models that we introduce in Chapter 2 are the only ordinal models we consider in this book.† However, we argue that it is useful to think of the less constrained models as either partially ordinal or ordinal in a less strict sense. The "partial" models that we discuss in Chapter 3 are partially ordinal in the sense that they maintain a strict stochastic ordering for a subset of variables in the model through the application of the parallel regression assumption, which constrains the effects of those variables to remain constant across outcome levels. We may consider an even broader class of models as ordinal if we only require that the models are not "permutation invariant" (McCullagh 1980, p. 116). Models designed for nominal dependent variables, such as multinomial logit, are permutation invariant, which means that arbitrary re-ordering of the outcome categories has no effect on the coefficients or on model fit (McCullagh 1980, p. 116). Permutation invariance is an important characteristic for nominal models because there is no inherent ordering to the categories. Using these criteria, a categorical model is "ordinal" if re-ordering the outcome categories leads to different results.

We refer to the models in this book as ordered regression models because they are variations on three ordinal models in the strictest sense (see Chapter 2). The variations are based on different ways to relax the parallel regression assumption. The results are not permutation invariant for almost all of the models we present in Chapters 2 through 4. The nonparallel adjacent category model is the only one that is permutation invariant, which means that it is ideally suited for nominal outcomes. The logit version of this model is equivalent to the well-known multinomial logit model (McFadden 1973). Although it is important to consider the defining characteristics of an ordinal model, the decision to label a model as "ordinal," "partially ordinal," or "nonordinal" is not as important as other modeling decisions, including the probability of interest and the application of the parallel regression assumption. Throughout the book, we focus on how the type of probability and application of the parallel regression assumption influence the results in ordered regression models. We refer to the sets of ordered regression models based on different probabilities of interest as "approaches to ordered regression models," and we organize Chapters 2 through 4 around these approaches.

The ordered regression models we consider are very useful for researchers examining ordinal outcomes for two main reasons. First, the models are appropriate for a wide

---

* The CDF describes the probability that a random variable, such as the error term, is less than or equal to a particular value (see Gill 2006, p. 348).

† We also consider several model extensions in Chapter 6, and the ones that retain the parallel regression assumption are also "ordinal" in the strictest sense.

variety of ordinal variables. Some ordinal outcomes have categories that represent intervals on an underlying latent continuous scale. For example, an ordinal measure of income with categories based on $25,000 intervals (e.g., <$25,000, $25,000 to <$50,000, $50,000 to <$75,000) is essentially a continuous variable grouped into intervals on an ordinal scale from lowest to highest income. As a result, scholars refer to some ordinal variables as "grouped continuous" variables (Anderson 1984, p. 2). However, other ordinal variables, such as educational attainment, are the result of a series of sequential decisions. The categories are ordered, but one must proceed through earlier stages in the sequence in order to reach a higher category. For example, in order to attain a postgraduate degree, one must first attain high school and college degrees. Finally, there are ordinal variables that have categories that are of substantive interest, such as subjective social class (Fullerton and Xu 2015), which are not the result of a sequential process and may or may not correspond with an underlying latent continuous variable. The broad range of models covered in this book is particularly important given the diversity of types of ordinal variables in the social and behavioral sciences.

Second, most of the models in this book allow for asymmetrical effects. Researchers often assume that the effects of independent variables are "symmetrical" (Jordan 1965; Lieberson 1985). In the context of public opinion research, the assumption of symmetry means that the effect of a variable on negative views toward an issue is equal in magnitude to its effect on positive views in the opposite direction. For example, if political conservatism is associated with increased opposition to same-sex marriage, then it should also be associated with decreased support for same-sex marriage, and the relative strength of each relationship should be equivalent.

However, the assumption of symmetry (Jordan 1965; Lieberson 1985) or bipolarity (Cacioppo and Berntson 1994; Cacioppo et al. 1997) may be concealing an asymmetrical relationship (Fullerton and Dixon 2010). Studies have shown that the racialization of welfare attitudes is greater for opposition than for support (Fullerton and Dixon 2009), and income reduces the incidence of low levels of life satisfaction but does not increase the incidence of high levels of life satisfaction (Boes and Winkelmann 2010). The partial and nonparallel models allow for asymmetrical relationships, whereas linear regression and traditional ordered models impose symmetry on the relationships (for more details, see Section 1.7).

1.2 Brief History of Binary and Ordered Regression Models

Ordered regression models are extensions of the binary regression model to ordinal outcomes. We will provide a brief history of the development of binary and ordered regression models before presenting the three major approaches to ordered regression models. For a more detailed history of binary and ordered regression models, see Greene and Hensher (2010, Ch. 4).

Binary regression models have their roots in the studies of toxicology and bioassay in the 1930s and 1940s (Bliss 1934a, 1934b; Finney 1971 [1947]). Bliss (1934a, 1934b) introduced the binary probit model as a method for analyzing the dosage mortality curve for nicotine spray and the percent of aphids killed. Bliss proposed the "method of probits" in order to model the S-shaped curve using Pearson's (1914) probability integral tables (Bliss 1934a, p. 38). The probit model is a nonlinear probability model that assumes the errors follow a normal distribution (Aldrich and Nelson 1984, pp. 34–37). Fisher (1935) was the first to suggest the use of maximum likelihood estimation (MLE) for the probit model, which Garwood (1941) and Finney (1947, 1952) developed further (Finney 1971, p. 52). Berkson (1944) introduced the binary logit model as an alternative to probit, which is a nonlinear probability model that assumes the errors follow a logistic distribution. Although he noted the tendency for logit and probit results to be very similar, Berkson (1951) made the case that logit models are preferable to probit models because logit is easier to fit and has better statistical properties. However, advances in statistical computing in the 1970s made the choice between logit and probit less important.

Aitchison and Silvey (1957) extended the binary probit model to ordinal outcomes with three or more categories with an application in entomology. Their study laid the groundwork for the cumulative probit model (see Chapter 2). Building on this earlier work, McKelvey and Zavoina (1975) were the first to present a formal treatment of the cumulative probit model with a theoretical motivation for the model, an iterative solution based on MLE, and an application to individual-level data with multiple independent variables. McKelvey and Zavoina (1975) used the idea of a latent continuous variable that was only measured at a few key points in the distribution. One observes an ordinal rather than continuous variable simply due to incomplete data (p. 105).

For example, occupational prestige is a continuous variable based on the perceived social standing of each occupation, but it may only be recorded in ordered intervals in a particular dataset. The observed variable is ordinal, but it is based on an underlying continuous distribution that cannot be directly observed. The idea of a latent continuous variable is often used as a motivation for ordered regression models (e.g., see Long 1997, pp. 116–122; Agresti 2010, pp. 53–55). However, it is not necessary for estimation or interpretation (McCullagh 1980, p. 110).

Snell (1964) similarly extended the logit model to the study of ordinal outcomes, and Walker and Duncan (1967) were the first to apply the logit model to ordinal outcomes using individual-level data. Although the roots of the cumulative logit model go back to Snell (1964), McCullagh (1980) is credited with developing the model, which he termed the "proportional odds" model due to the key model assumption. Until 1980, the development of regression models for ordinal outcomes was limited to cumulative probabilities. Fienberg (1980) and Goodman (1983) subsequently introduced the continuation ratio and adjacent category logit models, respectively. However, these latter two models remain relatively underutilized compared to the cumulative models in the social and behavioral sciences. The focus on cumulative models in the development of binary regression applications for ordinal outcomes has had a lasting effect because researchers continue to refer to the cumulative logit and probit models simply as "ordinal logit" and "ordinal probit."

1.3 Three Approaches to Ordered Regression Models

The so-called "ordinal logit" and "ordinal probit" models actually refer to sets of models based on the application of the binary logit or binary probit model to ordinal outcomes. We can separate these models into three major approaches: cumulative, stage, and adjacent. Cumulative models (e.g., McKelvey and Zavoina 1975; McCullagh 1980) have received the most attention in the literature and are used most often in practice. However, the stage and adjacent approach models offer unique advantages as well. The ordered regression models we consider in this book fit within a typology based on: (1) three approaches or probabilities of interest and (2) four applications of the parallel regression assumption within each approach (see Fullerton 2009).

The estimation of ordered regression models involves the creation of a series of $M - 1$ binary variables from the original, ordinal outcome variable with M categories. The equation for each new, binary outcome is referred to as a "cutpoint equation." The construction of the set of binary outcomes varies according to the probability of interest and therefore the type of approach (cumulative, stage, or adjacent). However, each method of constructing the binary outcomes recognizes the ordering of categories in the original variable. The binary regression models based on the $M - 1$ cutpoint equations are estimated simultaneously. The "cutpoints" are thresholds for the latent continuous variable and are typically treated as nuisance parameters given that they are sensitive to the particular identifying assumptions or parameterization of the model (see Long 1997, pp. 122–123). In the absence of a latent variable motivation, one may think of the cutpoints simply as constants in the $M - 1$ binary equations. In this book, we will introduce the models using the latent variable motivation as well as an alternative framework that presents the models as nonlinear probability models. Although these are two distinct frameworks for understanding ordered regression models, they are equivalent in terms of the statistical models. Therefore, the decision regarding whether or not to use a latent variable motivation for the model will not affect the results or substantive conclusions.

1.3.1 Cumulative Models

The probability of interest in cumulative approach models is the cumulative probability ($\Pr[y \leq m]$), which is the probability at or below a given outcome category, m (McCullagh 1980, p. 110). For example, if income is grouped into four intervals—1 = <\$25,000, 2 = \$25,000 to <\$50,000, 3 = \$50,000 to <\$75,000, and 4 = ≥\$75,000—the cumulative probability for category 2 is the probability that a respondent is located in one of the bottom two income categories (i.e., <\$50,000). Scholars have used cumulative models to analyze a broad range of ordinal outcomes, including worker attachment (Halaby 1986), health (House et al. 1994), and attitudes toward science (Gauchat 2011). For an outcome (y) with four categories (1–4), y is linked to the binary outcomes in the cutpoint equations ($y_1 - y_3$) by the measurement equations:

$$y_1 = \begin{cases} 1 \text{ if } y = 1 \\ 0 \text{ if } y = 2 - 4 \end{cases} \tag{1.1}$$

$$y_2 = \begin{cases} 1 \text{ if } y = 1 - 2 \\ 0 \text{ if } y = 3 - 4 \end{cases} \tag{1.2}$$

$$y_3 = \begin{cases} 1 \text{ if } y = 1 - 3 \\ 0 \text{ if } y = 4 \end{cases} \tag{1.3}$$

The sample size does not change across the cutpoint equations, which is a feature of cumulative models that is not shared by stage or adjacent models. For example, if there are three cutpoint equations, then every respondent is coded 0 or 1 for y_m in all three binary cutpoint equations. A respondent's value for y_m will vary across cutpoint equations, but the key point is that every respondent is included in each equation in the cumulative model. As a result, the binary equations are not independent, which has important implications for model estimation. We discuss this issue in more detail in Chapter 2. If the ordinal outcome is based on a sequence of stages with a logical starting point, then the conditional probability is more appropriate and one should consider using a stage model.

1.3.2 Stage Models

The probability of interest in stage models is the conditional probability ($\Pr[y = m | y \geq m]$), which is the probability for a particular category, m, given that the respondent is in that category or a higher one (Fienberg 1980, p. 110). As a result, respondents in lower categories are excluded. For example, if the dependent variable is a four-category measure of educational attainment—1 = less than high school, 2 = high school, 3 = some college, and 4 = college or more—the conditional probability for category 2 is the probability that a respondent has a high school degree given that he or she has a high school degree or more. We use Fullerton's (2009) terminology in referring to this as the "stage" approach because the cutpoint equations represent a series of stages. In order to reach a particular category, respondents must pass through each of the previous categories. Stage models are commonly referred to as "continuation ratio" models (Fienberg 1980; Long 1997; Agresti 2010; Tutz 2012) because the focus is on the sequential process of continuing from one stage to the next. However, they are also known as "stopping ratio" models because the conditional probability is the probability of stopping in a particular stage or outcome category (Yee 2015). We will refer to these models as continuation ratio models in order to be consistent with most studies in the literature.

Scholars have used stage models to examine ordinal variables such as educational attainment (Hauser and Andrew 2006), political participation (Fullerton and Stern 2013), vote overreporting (Fullerton et al. 2007), early literacy proficiency (O'Connell 2006), and smoking abstinence (Kendzor et al. 2012). However, stage models are not appropriate for every ordinal variable because there must be a logical starting point, respondents must pass through earlier stages to reach a later stage, and the stages must be irreversible (Tutz 1991). For instance, although one can estimate stage models for variables based on Likert scales, respondents do not have to disagree with a statement before they can agree with it.

For an outcome (y) with four categories (1–4), y is linked to the binary variables in the cutpoint equations ($y_1 - y_3$) by the measurement equations:

$$y_1 = \begin{cases} 1 \text{ if } y = 1 \\ 0 \text{ if } y = 2 - 4 \end{cases} \tag{1.4}$$

$$y_2 = \begin{cases} 1 \text{ if } y = 2 \quad | \; y \geq 2 \\ 0 \text{ if } y = 3 - 4 \quad | \; y \geq 2 \end{cases} \tag{1.5}$$

$$y_3 = \begin{cases} 1 \text{ if } y = 3 & | \ y \geq 3 \\ 0 \text{ if } y = 4 & | \ y \geq 3 \end{cases} \tag{1.6}$$

The outcome in the first cutpoint equation (y_1) is the same in the cumulative and stage models. However, the remaining cutpoint equations are different. In addition, the sample size becomes progressively smaller in later stages. Although there are several differences between the cumulative and stage models, neither approach focuses primarily on comparisons of individual categories. If the outcome categories are of substantive interest to the researcher, then the adjacent models may be more appropriate to use.

1.3.3 Adjacent Models

We refer to the probability of interest in adjacent models as the "adjacent probability" ($\Pr[y = m | y = m$ or $y = m + 1]$), which is the probability for a particular category, m, given that the respondent is in that category or the next highest one (Agresti 2010, p. 88). It is also a conditional probability, but the focus is limited to the two adjacent categories. Using the educational attainment example from the previous section, the adjacent probability for category 3 is the probability that a respondent has some college given that he or she has some college or a college degree.

Adjacent models are perhaps the most underutilized of the three sets of ordered regression models in the social and behavioral sciences. Cumulative models are by far the most popular, and scholars have utilized these models to analyze a variety of outcomes in several disciplines. Stage models are gaining popularity due to their applications in survival analysis (Albert and Chib 2001) and the fact that one may estimate nonparallel versions of these models as a series of separate binary regression models (see Fienberg 1980, pp. 110–111). However, adjacent models are better suited than cumulative models for certain types of ordinal variables, including Likert scales and other attitudinal scales that have "additional structure" (Sobel 1997, pp. 215–216).

One type of additional structure is the intensity of opinion on a topic. A Likert scale from strongly agree (1) to strongly disagree (5) measures the level of disagreement on an ordinal scale from 1 to 5, but the first and last two categories also measure the intensity of (dis)agreement. If the research question involves the degree of opinion polarization, then one may be more interested in the individual category probabilities than in the cumulative probabilities. In the context of logistic regression, one may be interested in the odds of strongly agree versus agree (1 vs. 2) rather than strongly agree versus agree to strongly disagree (1 vs. 2–5). The adjacent category model allows one to examine the intensity of agreement or disagreement (Sobel 1997, p. 215). In general, adjacent models are best suited for ordinal outcomes with categories that are of substantive interest to the researcher (Fullerton 2009, p. 320). Scholars have used adjacent models to analyze outcomes such as occupational mobility (Sobel et al. 1998), congressional incumbent approval (Jones and Sobel 2000), and health (Fullerton and Anderson 2013).

For an outcome (y) with four categories (1–4), y is linked to the binary variables in the cutpoint equations (y_1–y_3) by the measurement equations:

$$y_1 = \begin{cases} 1 \text{ if } y = 1 & | \ y = 1 \text{ or } 2 \\ 0 \text{ if } y = 2 & | \ y = 1 \text{ or } 2 \end{cases} \tag{1.7}$$

$$y_2 = \begin{cases} 1 \text{ if } y = 2 \mid y = 2 \text{ or } 3 \\ 0 \text{ if } y = 3 \mid y = 2 \text{ or } 3 \end{cases} \tag{1.8}$$

$$y_3 = \begin{cases} 1 \text{ if } y = 3 \mid y = 3 \text{ or } 4 \\ 0 \text{ if } y = 4 \mid y = 3 \text{ or } 4 \end{cases} \tag{1.9}$$

The outcome in the third cutpoint equation (y_3) is the same in the stage and adjacent models, but the first two cutpoint equations are different. Due to the focus on adjacent categories, no respondent is in more than two cutpoint equations regardless of the number of outcome categories.

The stage and adjacent models essentially select on the dependent variable because the samples are restricted in most or all of the cutpoint equations. For example, in stage models, the samples get progressively smaller in the cutpoint equations because respondents must "survive" to the next stage in order to be included in that equation. In adjacent models, only the respondents from two adjacent categories are included in any given cutpoint equation. In stage models, the progressively smaller sample sizes in later stages may lead to sample selection bias, and changes in coefficients across equations may reflect changes in unobserved heterogeneity rather than the "true effects" of the variables (Mare 2006, pp. 30–31). In adjacent models, the selection on the dependent variable requires the assumption of independence of irrelevant alternatives (IIA), which we discuss in Chapter 3. It is important to keep these limitations in mind when choosing an ordered regression model.

1.4 The Parallel Regression Assumption

The key assumption in the traditional versions of the cumulative, stage, and adjacent models is that the coefficients for each independent variable are constant across cutpoint equations, which is reflected in the following general equation for the linear predictor (η) in a traditional ordered regression model:

$$\eta = \tau_m - \mathbf{x}\boldsymbol{\beta} \quad (1 \le m < M), \tag{1.10}$$

where τ_m are "cutpoints" or constants from the binary cutpoint equation, m is an outcome category, $\mathbf{x}$ is a vector of independent variables, and $\boldsymbol{\beta}$ is a vector of coefficients. The cutpoint (τ) is the only parameter that varies across the binary cutpoint equations. The assumption of constant slopes is referred to as the "parallel regression" assumption because the probability curves are parallel.* For example, in a cumulative model with three cutpoint equations, the curve for $\Pr(y \le 1)$ is parallel to the curves for $\Pr(y \le 2)$ and $\Pr(y \le 3)$. The probability curves are shifted to the left or right by a constant, and the tangent lines for any given point in the probability curves are parallel (Long 1997, p. 141).

---

* In ordered logit models, it is also known as the "proportional odds" assumption because the log of the odds ratio is proportional to the change in the independent variable (Agresti 2010, p. 53).

The parallel regression assumption has several advantages. It maintains a strict stochastic ordering; it is required in order to assume a latent variable in cumulative models[*]; and, it allows the models to be relatively parsimonious. For an M-category ordinal outcome, there are $M - 1$ binary cutpoint equations and $M - 1$ coefficients for every independent variable if we completely relax the parallel regression assumption. Therefore, the added parsimony due to this assumption is evident as M increases. For example, if there are 10 independent variables, then the application of the parallel regression assumption for every independent variable results in 10 fewer parameters if $M = 3$, 20 fewer parameters if $M = 4$, and 30 fewer parameters if $M = 5$. In other words, there are 10 fewer parameters for every additional category. The loss of parsimony associated with relaxing the parallel regression assumption is even more important in small samples given the effect on the number of cases per parameter.

We focus on four different applications of the parallel regression assumption. Ordered regression models may apply the assumption to:

1. Every variable ("parallel")
2. A subset of variables ("unconstrained partial")
3. A subset of variables with additional constraints ("constrained partial")
4. None of the variables ("nonparallel")

Applications 2 through 4 represent extensions of the traditional ordered regression models. The unconstrained and constrained partial models differ according to the manner in which some coefficients vary across cutpoint equations.

1.5 A Typology of Ordered Regression Models

Fullerton (2009) developed a typology of ordered logistic regression models based on the probability of interest and the application of the parallel regression assumption. We use the same typology in this book, but we generalize it to include models with other distributional assumptions. In order to broaden the focus to include logit, probit, and complementary log log versions of the models, we use new terminology based on the two dimensions of Fullerton's (2009) typology without references to logit-specific model features, such as the odds. We present the generalized version of Fullerton's (2009) typology in Table 1.1.[†] The ordered regression models that we consider fit within a conceptual framework based on the approach to comparisons (cumulative, stage, or adjacent) and on the application of the parallel regression assumption to every independent variable, a subset of independent variables (unconstrained and constrained), or to no independent variables.

We refer to models that apply the assumption to every variable as the "parallel cumulative," "parallel continuation ratio," and "parallel adjacent category" models

---

[*] Relaxing the parallel regression assumption in cumulative models "detaches the observed outcomes from an underlying continuous process" (Greene and Hensher 2010, p. 190). Instead, one must treat the outcome as a "discrete multinomial random variable" (Greene and Hensher 2010, p. 192).

[†] We adapted this table from Fullerton (2009, p. 309, Table I).

TABLE 1.1

A Typology of Ordered Regression Models

Parallel Regression Assumption	Approach to Comparisons		
	Cumulative	Stage	Adjacent
Every independent variable	Parallel cumulative	Parallel continuation ratio	Parallel adjacent category
Some independent variables (unconstrained)	Unconstrained partial cumulative	Unconstrained partial continuation ratio	Unconstrained partial adjacent category
Some independent variables (constrained)	Constrained partial cumulative	Constrained partial continuation ratio	Constrained partial adjacent category
No independent variables	Nonparallel cumulative	Nonparallel continuation ratio	Nonparallel adjacent category

Source: Based on Fullerton, A. 2009. *Sociological Methods and Research* 38: 306–347.

(see the top row in Table 1.1). These are the traditional ordered regression models. They are the most parsimonious models, but the coefficients in parallel models have a greater potential for bias, given that the parallel regression assumption is often violated in practice.

The partial models represent the second and third applications of the parallel regression assumption. The "unconstrained partial cumulative," "unconstrained partial continuation ratio," and "unconstrained partial adjacent category" models allow the coefficients for a subset of independent variables to freely vary across cutpoint equations (see the second row in Table 1.1). The manner in which the coefficients vary across equations is not determined a priori. On the other hand, the "constrained partial cumulative," "constrained partial continuation ratio," and "constrained partial adjacent category" models impose one or more constraints on the pattern of coefficient variation across cutpoint equations (see the third row in Table 1.1). If there are only two cutpoint equations, then the coefficients for a group of independent variables will change by the same factor across equations. For instance, the effects may be 20% weaker in the second cutpoint equation than in the first cutpoint equation. Fullerton and Xu (2012) provide a conceptual justification for the constrained partial cumulative model and a new estimation procedure that improves upon an earlier formulation for constrained partial models (Hauser and Andrew 2006).

The last group of ordered regression models relaxes the parallel regression assumption for every independent variable. We refer to these models as the "nonparallel cumulative," "nonparallel continuation ratio," and "nonparallel adjacent category" models (see the bottom row in Table 1.1). The nonparallel models are the most flexible but also the least parsimonious models. Using the logit versions, the nonparallel adjacent model is equivalent to the well-known multinomial logit model (Long 1997, p. 146), and the nonparallel cumulative and stage models are more commonly known as "generalized ordered logit" (Williams 2006) and "sequential logit" (Tutz 1991), respectively. Although multinomial logit is designed for nominal dependent variables, it is a flexible alternative that may be appropriate if the outcome is well suited for adjacent models but the parallel regression assumption is violated. It allows for ordinal patterns to emerge in the data, but it does not formally impose ordinality on the relationships (Long 2015).

1.6 Link Functions

In this book, we focus on logit, probit, and complementary log log (cloglog) versions of ordered regression models. Using a generalized linear modeling framework (McCullagh and Nelder 1989), these three model types represent different choices of "link functions" ($g[\mu]$) that relate the linear predictor (η) to the expected value or probability of interest (μ):

$$\eta = g(\mu) \tag{1.11}$$

In order to estimate the model, it is necessary to assume that the outcome is the realization of a random variable based on a probability distribution. To specify a model, one must select a particular probability distribution for the random component and a link function relating the linear predictor to the expected value (see McCullagh and Nelder 1989, pp. 26–32). Ordered logit models assume a logistic distribution for the random component and use a logit link function:

$$\ln\left(\frac{\mu}{1-\mu}\right) = \eta = \tau_m - \mathbf{x}\boldsymbol{\beta} \tag{1.12}$$

The linear predictor in the ordered logit model is the log of the odds: $\ln\left(\frac{P}{1-P}\right)$. The exponent of the linear predictor is the odds: $\frac{P}{1-P}$. Therefore, the parallel cumulative logit model is also referred to as the cumulative odds or proportional odds model (McCullagh 1980). Ordered probit models assume a normal distribution and use the inverse normal link function:

$$\Phi^{-1}(\mu) = \eta = \tau_m - \mathbf{x}\boldsymbol{\beta} \tag{1.13}$$

where $\Phi(\mu)$ is the normal CDF. Finally, ordered complementary log log models assume an extreme value distribution and use the complementary log log link function:

$$\ln\{-\ln(1-\mu)\} = \eta = \tau_m - \mathbf{x}\boldsymbol{\beta} \tag{1.14}$$

These equations refer to the parallel versions of the models. We focus on these three link functions because they are the most commonly used links for ordered regression models. Ordered logit and probit have symmetrical CDFs, which means that the estimates are unaffected by a reversal of outcome categories. In other words, the models display "palindromic invariance" (McCullagh 1980, p. 116). It does not matter if the categories are ranked from high to low or low to high. Reversing the category order will change the sign on the slopes, but it will not affect the significance tests or model fit. However, the CDF for the complementary log log model is asymmetric, and reversing the category order changes the model estimates. At the lower end of the probability scale, the logistic and complementary log log functions are very similar, but at the upper end the complementary log log function approaches infinity more slowly than the other two functions (McCullagh and Nelder 1989, p. 109). For ordinal outcomes that are appropriate for stage models, reversing the category order would be nonsensical because there is a natural starting point.

For example, the educational attainment process obviously begins with "less than high school" rather than a "college degree or more." Therefore, the lack of palindromic invariance is not a problem for stage models using the complementary log log link function.

1.7 Asymmetrical Relationships in Partial and Nonparallel Models

In addition to simplifying the model, the parallel regression assumption imposes symmetry on the effects of independent variables. The effects of variables are constant across cutpoint equations, which means that if a variable increases strong disagreement with a particular statement it will have an equally strong effect in the opposite direction for strong agreement. For example, if we have an ordinal outcome with five categories (1–5) and a parallel cumulative logit model, then the effect of a variable on the log odds of 1 versus 2–5 will be equal to the effect on the log odds of 1–4 versus 5. If categories 1 and 5 represent strong opposition and strong support, respectively, then the parallel assumption constrains the coefficients so that the effects on strong opposition and lack of strong support are equivalent. This may only be a reasonable assumption for some variables in the model. Partial and nonparallel models allow the researcher to relax the parallel assumption for one or more variables. As a result, asymmetrical patterns may emerge.[*]

Ideally, one should rely on theory to decide whether or not it is reasonable to expect asymmetrical relationships (e.g., see Fullerton and Dixon 2009). However, theories in the social and behavioral sciences often do not suggest whether the relationships should be symmetrical or asymmetrical or whether the coefficients should be constant or variable across outcome categories. In that case, researchers may rely on empirical tests of the parallel regression assumption. However, one must be careful not to over-fit the data (Williams 2006; Long 2015). Williams (2015) discusses several applications of partial and nonparallel models and alternative ways to interpret the results.

1.8 Hypothesis Testing and Model Fit in Ordered Regression Models

Although our focus is on interpretation in the empirical examples, we will also address statistical significance and model fit. There are a variety of measures available for hypothesis testing, including z-tests, Wald tests, and likelihood ratio (LR) tests. We primarily use z-tests for the β's. In order to determine whether the raw coefficient is statistically significant, one typically tests the following null hypothesis: $H_0 : \beta_k = 0$. The equation for the test statistic, z, is:

$$z = \frac{\hat{\beta}_k}{S_{\hat{\beta}_k}}$$

(1.15)

---

[*] In a strict sense, the presence of symmetrical effects suggests that the effect of x on the log odds of 1 versus 2–5 should be equal to the effect of x on the log odds of 5 versus 1–4 (multiplied by –1). For example, if the coefficient for x in the first equation (1 vs. 2–5) is 0.5, then the coefficient for x in the second equation (5 vs. 1–4) should be –0.5. However, the parallel regression assumption only guarantees symmetrical coefficients in this strictest sense in the logit and probit versions of ordered regression models. The cloglog distribution is asymmetrical.

where $\hat{\beta}_k$ is the estimate of the coefficient for x_k from the ordered regression model, and $S_{\hat{\beta}_k}$ is the standard error of that estimate. If the probability or "*p*-value" associated with z is below the critical value or "alpha level" (e.g., $\alpha = 0.05$), then there is sufficient evidence to reject the null hypothesis and conclude that the effect of x_k is statistically significant. It is also possible to test the statistical significance of $\hat{\beta}_k$ using Wald and LR tests. We discuss these tests in Chapter 5 in the context of testing the parallel regression assumption. It is important to keep in mind that statistical significance does not guarantee substantive significance. In large samples, it is common for variables to have statistically significant coefficients even if their effects on the predicted probabilities are very small. We will return to this distinction between statistical and substantive significance when comparing partial and nonparallel models in Chapters 3 and 4, respectively, and when testing the parallel regression assumption in Chapter 5.

There are also several different measures of model fit for ordered regression models. We will present a few pseudo-R^2 measures in the empirical examples and consider other measures such as the Akaike information criterion (AIC) and Bayesian information criterion (BIC) in Chapter 5. We focus on three pseudo-R^2 measures: the well-known McFadden R^2 (McFadden 1973), a recently developed measure based on ordinal dispersion (Lacy 2006), and a measure based on nominal dispersion that is equivalent to Goodman and Kruskal's tau (Haberman 1982). McFadden (1973, p. 121) developed a fit statistic based on the relative change in the log-likelihood function from the null model to the current model:

$$R_M^2 = 1 - \frac{LL_C}{LL_N} \qquad (1.16)$$

where LL_C is the log likelihood for the current model, and LL_N is the log likelihood for the null model. The measure was intended to be analogous to the R^2, which means that LL_N is a measure of the total sum of squares and LL_C is a measure of the residual sum of squares. However, one could also think of it as the proportional reduction in deviance (−2LL) from the null model to the current model. We include this measure because of its widespread use in the social and behavioral sciences.[*]

Despite the advantages of R_M^2, it has been criticized on the basis of a lack of interpretability (Long 1997) and on its performance in simulation studies (Lacy 2006). Additionally, it does not take into account the ordinal nature of y. Therefore, we also include Lacy's (2006) recently developed pseudo-R^2 measure based on "ordinal dispersion" (Blair and Lacy 2000), which is a measure of the degree of diversity in the distribution of an ordinal variable. Ordinal dispersion is a function of the cumulative relative frequency distribution (Lacy 2006, p. 473):

$$d^2 = \sum_{j=1}^{k-1} F_j \left(1 - F_j\right) \qquad (1.17)$$

where k is the number of categories and F_j is the cumulative proportion for category j. This new measure of model fit, R_O^2, is the proportion of the total amount of ordinal dispersion that is explained by the ordered regression model (Lacy 2006, p. 475):

$$R_O^2 = 1 - \frac{E_x\left[d^2(\mathbf{x})\right]}{d^2} \qquad (1.18)$$

---

[*] In Stata, the McFadden R^2 is the "pseudo R^2" value reported in the default output for regular commands (e.g., ologit) and user-written commands [e.g., ocratio (Wolfe 1998)].

where $E_x[d^2(\mathbf{x})]$ is the conditional ordinal dispersion based on cumulative predicted probabilities, and d^2 is the unconditional ordinal dispersion based on the marginal distribution (for more details, see Lacy 2006, pp. 475–476). The values for R_O^2 range from 0 (no ordinal variation explained) to 1 (all of the ordinal variation explained). For a discussion of alternative ordinal measures of model fit, see Lacy (2006, pp. 480–484).

Finally, we also include a pseudo-R^2 measure based on nominal dispersion. This will be a particularly useful measure when we consider models that are essentially hybrids of ordinal and nominal models, such as the partial and nonparallel models in Chapters 3 and 4, respectively. The nominal dispersion R^2 measure is based on Simpson's Index of Diversity (Agresti and Agresti 1978, p. 206):

$$D = 1 - \sum_{j=1}^{j=k} p_j^2 \qquad (1.19)$$

where p_j is the proportion for category j and k is the number of categories. The nominal R^2 is the proportion of nominal dispersion explained by the nominal or ordered regression model:

$$R_N^2 = 1 - \frac{E_x\left[D(\mathbf{x})\right]}{D} \qquad (1.20)$$

where $E_x[D(\mathbf{x})]$ is the conditional nominal dispersion based on predicted probabilities, and D is the unconditional nominal dispersion based on the marginal distribution.[*] The values for R_N^2 range from 0 (no nominal variation explained) to 1 (all of the nominal variation explained).

1.9 Datasets Used in the Empirical Examples

1.9.1 Cumulative Models of Self-Rated Health

We will illustrate the cumulative approach models using a general self-rated health example from the 2010 GSS.[†] Self-rated health is an overall measure of perceived health with the following ordered categories: poor, fair, good, and excellent. For this example, we use data from the 2010 GSS. The GSS is a multistage, stratified sample of the noninstitutionalized, English-speaking adult population living in the United States (Smith et al. 2013). After dropping cases with missing data, we have a sample size of 1,129. We present descriptive statistics for the dependent and independent variables in Table 1.2. Most respondents described their health as either excellent (25.16%) or good (46.59%). Fewer respondents claimed that their health was fair (22.67%) or poor (5.58%). We model general self-rated health using several independent variables, including age (in years), age-squared, female (sex: 1 = female, 0 = male), race/ethnicity (binary variables for White [ref.], Black, Latino, and Other Race/Ethnicity), education (in years), employed

---

[*] We thank Michael Lacy for suggesting the inclusion of this measure and modifying his Stata routine ("r2o") to include it in addition to the ordinal dispersion measure.

[†] We use three main examples in Chapters 2 through 5, which we describe in this section. We use a few additional examples for the model extensions, which we describe in Chapter 6.

TABLE 1.2

Descriptive Statistics and Descriptions for Variables in Models of Self-Rated Health

Dependent Variable

General Self-Rated Health

	Pct
1 = Poor	5.58
2 = Fair	22.67
3 = Good	46.59
4 = Excellent	25.16

Independent Variables

	Mean	SD	Range	Description
Age	46.89	16.55	18–89	Age in years
Age-squared	2472.43	1677.34	324–7921	Age-squared
Female	0.54	0.50	0–1	Sex: 1 = female, 0 = male
Race/ethnicity				
White (ref.)	0.71	0.46	0–1	Race: 1 = white, 0 = else
Black	0.14	0.34	0–1	Race: 1 = black, 0 = else
Latino	0.12	0.32	0–1	Race: 1 = Latino, 0 = else
Other race/ethnicity	0.04	0.20	0–1	Race: 1 = other race, 0 = else
Education	13.69	3.07	0–20	Years of schooling
Employed	0.60	0.49	0–1	1 = employed, 0 = else
Log income	9.84	1.19	5.56–11.69	Natural log of household income
Married	0.46	0.50	0–1	Marital status: 1 = married, 0 = else

Note: N = 1,129. Data are from the 2010 General Social Survey.

(1 = yes, 0 = no), the natural log of household income, and married (marital status: 1 = married, 0 = else).

1.9.2 Stage Models of Educational Attainment

The main example we will use for stage or continuation ratio models is educational attainment, which is a popular method for this outcome given its sequential nature (e.g., Mare 1980; Hauser and Andrew 2006). Educational attainment is an indicator of the highest degree achieved with the following ordered categories: less than high school, high school or equivalent, associate or junior college, college, and postgraduate.* We use data from the 2010 GSS in order to examine the factors that influence educational attainment, and we limit the sample to adults 25 and older (N = 1,286). We present descriptive statistics for the dependent and independent variables in Table 1.3. According to these descriptive results, 9.8% of respondents have less than a high school degree, 44.32% have a high school degree, 7.47% have an associate's degree, 24.34% have a 4-year college degree, and 14.07% have a postgraduate degree. We control for several socio-demographic independent variables in the models, including age (in years), female (sex: 1 = female, 0 = male), and race/ethnicity

---

* Technically, respondents do not need to pass through the junior college stage. We will assume that they do in order to use the continuation ratio models for educational attainment. Alternatively, one could combine the high school and junior college categories or include respondents with "some college" in the junior college category.

TABLE 1.3

Descriptive Statistics and Descriptions for Variables in Models of
Educational Attainment

Dependent Variable

Educational Attainment

	Pct
1 = Less than high school	9.80
2 = High school	44.32
3 = Associate	7.47
4 = College	24.34
5 = Postgraduate	14.07

Independent Variables

	Mean	SD	Range	Description
Age	50.54	15.95	25–89	Age in years
Female	0.56	0.50	0–1	Sex: 1 = female, 0 = male
Race/ethnicity				
White (ref.)	0.77	0.42	0–1	Race: 1 = white, 0 = else
Black	0.09	0.29	0–1	Race: 1 = black, 0 = else
Latino	0.10	0.31	0–1	Race: 1 = Latino, 0 = else
Other race/ethnicity	0.04	0.19	0–1	Race: 1 = other race, 0 = else
Parents' education				
Mother's education	11.54	3.76	0–20	Mother's years of schooling
Father's education	11.49	4.24	0–20	Father's years of schooling

Note: $N = 1,286$. Data are from the 2010 General Social Survey.

(binary variables for White [ref.], Black, Latino, and Other Race/Ethnicity). We also include two measures of family background based on previous research: mother's and father's education (in years).

1.9.3 Adjacent Models of Welfare Attitudes

Finally, for the adjacent category models, we will use welfare spending attitudes as our example. Adjacent category models are well-suited for government spending questions because the categories typically reflect distinct spending views, but the assumption of a sequential decision-making process is not applicable. The data for this example come from the 2012 GSS. The dependent variable, welfare spending attitude, is a three-category ordinal measure with the following categories: "too much," "about right," and "too little." The independent variables are age (in years), female (sex: 1 = female, 0 = male), log family income, education (in years), ideology (1 = extremely liberal, 4 = moderate, 7 = extremely conservative), party identification (binary variables for democrat, independent, and republican [ref.]), and racial attitude (1 = Blacks are hardworking, 7 = Blacks are lazy). In accordance with previous research (Gilens 1999), we limit the sample to White respondents ($N = 392$), given the racialized nature of welfare attitudes in the United States. We present descriptive statistics for the dependent and independent variables in Table 1.4. According to these results, 49% of the White respondents say that we spend too much, 33% say that we spend about the right amount, and 18% say that we spend too little on welfare.

TABLE 1.4

Descriptive Statistics and Descriptions for Variables in Models of Welfare Attitudes

Dependent Variable

Attitudes Toward Welfare Spending

	Pct
1 = Too much	48.98
2 = About right	33.42
3 = Too little	17.60

Independent Variables

	Mean	SD	Range	Description
Age	46.93	17.05	18–87	Age in years
Female	0.61	0.49	0–1	Sex: 1 = female, 0 = male
Income	9.98	1.14	5.50–11.95	Natural log of household income
Education	13.91	2.77	2–20	Years of schooling
Ideology	3.97	1.45	1–7	1 = extremely liberal, 4 = moderate, 7 = extremely conservative
Party identification				
Republican (ref.)	0.27	0.44	0–1	1 = Republican, 0 = else
Democrat	0.31	0.46	0–1	1 = Democrat, 0 = else
Independent	0.42	0.49	0–1	1 = Independent, 0 = else
Racial attitude	4.22	1.01	1–7	Perception of Blacks' work ethic: 1 = hardworking, 7 = lazy

Note: N = 392. Data are from the 2012 General Social Survey.

1.10 Example: Education and Welfare Attitudes

A brief example focusing on the relationship between education and attitudes toward welfare spending will help illustrate the differences among the three groups of models and the importance of using ordered regression models rather than linear or binary models for ordinal outcomes. There are three GSS categories for measuring welfare attitudes. Therefore, there are two cutpoint equations in each ordered regression model, but the models vary in terms of the categories included in each equation. In a cumulative logit model, the cutpoint equations would be

$$\ln\left(\frac{\Pr[y = \text{too much} \mid \mathbf{x}]}{\Pr[y = \text{about right or too little} \mid \mathbf{x}]}\right) \tag{1.21}$$

$$\ln\left(\frac{\Pr[y = \text{too much or about right} \mid \mathbf{x}]}{\Pr[y = \text{too little} \mid \mathbf{x}]}\right) \tag{1.22}$$

where $\mathbf{x}$ is a vector of independent variables. Each welfare attitudes category is included in the two cutpoint equations. However, in a continuation ratio logit model, the sample would be smaller in the second cutpoint equation:

$$\ln\left(\frac{\Pr[y = \text{too much} \mid \mathbf{x}]}{\Pr[y = \text{about right or too little} \mid \mathbf{x}]}\right) \tag{1.23}$$

$$\ln\left(\frac{\Pr[y=\text{about right}\mid\mathbf{x}]}{\Pr[y=\text{too little}\mid\mathbf{x}]}\right) \tag{1.24}$$

The continuation ratio logit model would not be appropriate for welfare attitudes because one does not have to oppose welfare spending before supporting it. The adjacent category logit model does not require this type of sequential decision-making process. It simply focuses on the local or adjacent comparisons:

$$\ln\left(\frac{\Pr[y=\text{too much}\mid\mathbf{x}]}{\Pr[y=\text{about right}\mid\mathbf{x}]}\right) \tag{1.25}$$

$$\ln\left(\frac{\Pr[y=\text{about right}\mid\mathbf{x}]}{\Pr[y=\text{too little}\mid\mathbf{x}]}\right) \tag{1.26}$$

The cumulative logit model is often used to examine attitudes toward government spending, but the adjacent category logit model would be an appropriate model to use as well. The adjacent category model may actually be the ideal model to use if the categories reflect three distinct spending views rather than points along a latent, continuous distribution. In order to allow for asymmetrical determinants of welfare attitudes, one could use a partial or nonparallel model.

Some researchers are hesitant to use ordered regression models to analyze ordinal outcomes such as welfare attitudes because these models are seen as too complicated or difficult to interpret. One of our goals in this book is to show that these models are simply extensions of the familiar binary regression model and are very similar in terms of the ease of use and interpretation. In this introductory example, we will also show that it is important to use ordered regression models for ordinal outcome variables because the use of other models, such as linear regression or binary regression, may produce misleading results.

In Table 1.5, we present the results from several models of welfare attitudes with education as the only independent variable. Although the use of ordered regression models has become more widespread in recent years, some researchers continue to analyze ordinal outcomes using linear regression. According to the linear regression results in Table 1.5, education is statistically unrelated to welfare attitudes because the z-test is not significant. The results from a parallel cumulative logit or "proportional odds" model lead one to the same conclusion. The coefficients in both models are negative and nonsignificant, which

TABLE 1.5

Coefficients from Linear Regression, Parallel Cumulative Logit, and Binary Logit Models of Welfare Attitudes

	Linear Regression	Parallel Cumulative Logit	Binary Logit (Y>1)	Binary Logit (Y>2)
Education	−0.013	−0.023	0.004	−0.098*
	(−0.949)	(−0.666)	(0.123)	(−2.028)

Note: N = 392. Numbers in parentheses are z-ratios. Data are from the 2012 General Social Survey.
*p < 0.05, **p < 0.01, ***p < 0.001.

means that more educated respondents are somewhat less supportive of welfare spending, but that difference is not significantly different from zero. Therefore, one would conclude that education is not an important predictor of welfare attitudes.

Some researchers justify the use of linear regression for ordinal outcomes by noting that the results are very similar to those from the proportional odds model. However, in this case, one could argue that linear regression and parallel cumulative logit provide a similarly poor fit to the data. This is clear when we examine the results from binary logit models based on two different dichotomizations of the ordinal measure (2–3 vs. 1 and 3 vs. 1–2). Education is not significantly associated with a lack of opposition to welfare spending ("about right" or "too little" vs. "too much"), but it has a significant, negative association with support for welfare spending ("too little" vs. "too much" or "about right"). In other words, an increase in education is not associated with an increase in opposition to welfare spending, but it is associated with a decrease in support for it. These results suggest that welfare support is not the exact opposite of welfare opposition with respect to years of education. The decrease in support is offset by an increase in mixed support, which we examine in greater detail in the examples in Chapters 2 through 5. This is an asymmetrical pattern that is clearly not detectable using the traditional linear regression model or a parallel ordered regression model.

The results in this example highlight a potential problem with another common approach to analyzing ordinal outcomes. Dichotomizing the outcome in order to use binary logit or probit requires one to focus on just one cutpoint equation. However, it is not always clear which equation one should select. In this example, one would conclude that education is not related to welfare attitudes based on the first cutpoint equation, but one would come to the opposite conclusion based on the second equation. Fortunately, ordered regression models do not require one to make this choice. These models utilize every cutpoint equation. Additionally, the partial and nonparallel models allow one to uncover asymmetrical patterns, such as this one.

1.11 Organization of the Book

Chapter 2 presents the three "parallel" ordered regression models. These are the traditional ordered regression models, which impose the parallel regression assumption for every independent variable in the model. We introduce the cumulative and continuation ratio models using a latent variable framework, but we also provide an alternative framework for understanding them as nonlinear probability models that does not require the assumption of a latent continuous variable. The alternative framework is particularly useful because adjacent models do not rely on a latent variable motivation, and the latent variable assumption no longer holds for cumulative models if the parallel regression assumption is relaxed for one or more independent variables. One could apply the latent variable motivation to adjacent models given that these models are constrained versions of the multinomial logit model, which one may interpret within a random utility framework (McFadden 1973). However, these models are typically presented as nonlinear adjacent probability models without any reference to an underlying latent variable (e.g., Clogg and Shihadeh 1994, pp. 147–149; Long 1997, p. 146; Powers and Xie 2008, pp. 226–227; Agresti 2010, pp. 88–96).

Chapter 3 introduces the unconstrained and constrained partial models, which relax the parallel regression assumption for a subset of independent variables. These models are extensions of the traditional ordered regression models. We also discuss the connections between these models and the "stereotype ordered regression model" (Anderson 1984). Chapter 4 presents the nonparallel models, which are extensions of the traditional models that relax the parallel regression assumption for every independent variable in the model. These models are the most flexible but also the least parsimonious. Chapter 5 reviews tests for the parallel regression assumption in the extant literature and proposes new variations on several of these tests. We also discuss important practical concerns related to tests of the parallel regression assumption, including the use of exploration and verification sub-samples to avoid "over-fitting" the data (Long 2015, p. 199), the trade-off between variance and bias (see Agresti 2010, p. 76), and the choice of a reference model for comparisons. Chapter 6 reviews several extensions of ordered regression models, including heterogeneous choice models, multilevel ordered models, and the Bayesian approach to ordered regression models. We provide brief examples using Stata and R in the appendices of Chapters 2 through 4. Additional examples, maximum likelihood routines, and syntax to reproduce the results in this book are available online.[*]

---

[*] http://sociology.okstate.edu/faculty-staff-directory/faculty-directory/dr-andrew-fullerton

2

Parallel Models

Parallel models are ordered regression models that impose the parallel regression constraint for every independent variable in the model. In other words, the slopes do not vary across cutpoint equations. As a result, parallel models are "ordinal" models in the strictest sense of the term because the parallel regression assumption ensures a strict stochastic ordering (McCullagh 1980, pp. 115–116), which we discussed in detail in Chapter 1. Parallel models are also the most parsimonious ordered regression models covered in this book. Parallel models only require the estimation of one coefficient for each independent variable and $M–1$ cutpoints or constants, where M is the number of outcome categories. For example, a parallel model for an outcome with 4 categories and 10 independent variables requires the estimation of 13 parameters: 10 slopes and 3 cutpoints. By contrast, partial and nonparallel models with the same set of variables would require the estimation of as many as 33 parameters. Parsimony is one of the most important advantages of parallel models, which is particularly important for analyses based on small samples.

There are three different types of parallel models: cumulative, stage, and adjacent. These models differ from one another based on the probability of interest. The type of parallel model one should use depends on several different factors, including the data generating process, the meaningfulness of the individual categories, and the reversibility of outcome categories. We consider all three issues as we introduce the three parallel ordered regression models. In this chapter, we discuss the three parallel models, methods of interpretation, and the estimation of parallel models. In addition, we illustrate the different features of each model with empirical examples.

2.1 Parallel Cumulative Model

The parallel cumulative model is an ordered regression model that imposes the parallel regression assumption for every independent variable and focuses on the cumulative probability (i.e., $\Pr[y \leq m|\mathbf{x}]$) in each cutpoint equation. As we discussed in Chapter 1, the original, ordinal dependent variable, y, is recoded into a set of binary outcomes corresponding to different points in the cumulative distribution of y. The idea of a latent continuous variable underlying the observed ordinal outcome is a commonly used framework for understanding the parallel cumulative model, and we describe this framework in the next section. However, the latent variable motivation is not necessary for model estimation or interpretation (McCullagh 1980, p. 110). Therefore, we also present the parallel cumulative model as a nonlinear cumulative probability model without any reference to a latent variable. The models with and without the latent variable assumption are equivalent, which means that the decision to use this motivation for the model will not affect the results. After introducing the model using both frameworks, we will consider an example from the field of health research.

2.1.1 A Latent Variable Model

McKelvey and Zavoina (1975) were among the first to introduce the parallel cumulative probit model using a latent variable framework. They made the explicit assumption that the ordinal dependent variable is actually a continuous variable measured on an ordinal scale due to incomplete data (McKelvey and Zavoina 1975, p. 105). For example, income is a continuous variable that is often measured on an ordinal scale in surveys because it is easier to determine a respondent's income interval (e.g., $40,000 to < $60,000) than his or her exact income. One can also use a latent variable framework for other types of ordinal dependent variables, such as Likert scales. Respondents may have an underlying propensity to agree or disagree with a particular statement. This latent propensity is continuous but only observed at a few key points in the underlying distribution (e.g., strongly agree, agree, disagree, or strongly disagree). The structural model for the latent, continuous outcome, y^*, is

$$y^* = \mathbf{x}\boldsymbol{\beta} + e \tag{2.1}$$

where $\mathbf{x}$ is a vector of independent variables, $\boldsymbol{\beta}$ is a vector of coefficients, and e is the error term. Focusing on the cumulative distribution, the latent continuous outcome is linked to the observed ordinal outcome, y, through the following measurement equation:

$$y \le m \text{ if } y^* < \tau_m \quad (1 \le m \le M) \tag{2.2}$$

where m is an outcome category and τ_m are "cutpoints" or "thresholds" on the latent continuous scale. The categories for y correspond to specific intervals of y^* bounded by cutpoints. For category m, the lower bound of the interval is τ_{m-1} and the upper bound is τ_m. In other words,

$$y = m \text{ if } \tau_{m-1} \le y^* < \tau_m \quad (1 \le m \le M) \tag{2.3}$$

For example, $y = 2$ if y^* is between τ_1 and τ_2. Theoretically, the range for y^* is from $-\infty$ to ∞. Therefore, we assume that $\tau_0 = -\infty$ and $\tau_M = \infty$. The remaining cutpoints are treated as unknown and are estimated in the model along with the slopes. Based on y^*, the equation for the cumulative probability is

$$\Pr(y \le m | \mathbf{x}) = \Pr(y^* < \tau_m | \mathbf{x}) \tag{2.4}$$

Substituting $\mathbf{x}\boldsymbol{\beta} + e$ for y^*,

$$\Pr(y \le m | \mathbf{x}) = \Pr(\mathbf{x}\boldsymbol{\beta} + e < \tau_m | \mathbf{x}) \tag{2.5}$$

Then, subtracting $\mathbf{x}\boldsymbol{\beta}$ from both sides of the inequality,

$$\Pr(y \le m | \mathbf{x}) = \Pr(e < \tau_m - \mathbf{x}\boldsymbol{\beta} | \mathbf{x}) \tag{2.6}$$

We will consider three versions of the parallel cumulative model based on different assumptions about the distribution of e. Regardless, we may express the cumulative probability as a function of τ_m and $\mathbf{x}\boldsymbol{\beta}$ (see Long 1997, p. 121):

$$\Pr(y \le m | \mathbf{x}) = \Pr(y^* < \tau_m | \mathbf{x}) = F(\tau_m - \mathbf{x}\boldsymbol{\beta}) \tag{2.7}$$

where F is the cumulative distribution function (CDF). The probability for a particular category, m, is the difference between the cumulative probabilities for categories m and $m-1$:

$$\Pr(y = m|\mathbf{x}) = \Pr(y \le m|\mathbf{x}) - \Pr(y \le m-1|\mathbf{x}) \tag{2.8}$$

or

$$\Pr(y = m|\mathbf{x}) = F(\tau_m - \mathbf{x}\boldsymbol{\beta}) - F(\tau_{m-1} - \mathbf{x}\boldsymbol{\beta}) \tag{2.9}$$

The parallel cumulative logit model assumes a logistic error distribution, whereas the parallel cumulative probit and parallel cumulative complementary log log models assume normal and extreme value distributions for the errors, respectively. The ordered nature of the cutpoints guarantees that the predicted probabilities will be nonnegative (Tutz 2012, p. 244). This is a feature of the parallel cumulative model that is not shared by the parallel continuation ratio or by parallel adjacent category models.

Another unique feature of the parallel cumulative model is the collapsibility of adjacent outcome categories (Agresti 2010, p. 56). The values for $\boldsymbol{\beta}$ remain the same regardless of the number of τ's. For example, collapsing two categories for educational attainment, such as "some college" and "college," results in one less cutpoint but in identical estimates for the slopes. The β's are invariant to the choice of the number of outcome categories. Additionally, if categories are collapsed to the point that $M = 2$, then the parallel cumulative model is equivalent to the binary regression model.

Finally, in order to identify the parallel cumulative model, one additional assumption is necessary. The distributional assumptions about the error term allow us to identify the variance of the latent dependent variable, y^*. However, the mean of y^* is not identified because one can add an arbitrary constant to the cutpoints and the constant without affecting the predicted probabilities (Long 1997, pp. 122–123). In order to identify the model, one typically assumes that either $\tau_1 = 0$ or $\beta_0 = 0$. Programs vary in terms of which assumption they use to parameterize the model. For example, SAS uses the former assumption, whereas Stata uses the latter. The parameterization choice is arbitrary and only affects the constant and cutpoints.

2.1.2 A Nonlinear Cumulative Probability Model

The latent variable framework is often used as a motivation for the parallel cumulative model. Although the assumption of a latent continuous variable is convenient and provides a theoretical rationale for the model, it is not always a realistic assumption and it is not necessary for estimation or interpretation. The parallel cumulative model is a series of binary regression models with a common set of parameters for the independent variables (Tutz 2012, p. 44). The latent variable framework is also used as a motivation for the binary regression model, but one may alternatively think of the binary regression model as a nonlinear probability model (Aldrich and Nelson 1984, pp. 30–35). Similarly, one may view the parallel cumulative model as a nonlinear cumulative probability model (Long and Freese 2014, p. 314). The general form of the equation for the nonlinear cumulative probability model is:

$$\Pr(y \le m|\mathbf{x}) = F(\tau_m - \mathbf{x}\boldsymbol{\beta}) \tag{2.10}$$

where F typically represents the CDF for the logit, probit, or complementary log model. This is the same equation that we previously derived within the latent variable framework.

As we discussed in Chapter 1, the ordinal dependent variable is recoded into a set of binary outcomes in the cutpoint equations. If $M = 4$, then the parallel cumulative model estimates the following three binary cutpoint equations simultaneously based on different points in the cumulative distribution for y:

$$\text{Cutpoint equation \#1: } \Pr(y \leq 1|\mathbf{x}) = F(\tau_1 - \mathbf{x}\boldsymbol{\beta})$$

$$\text{Cutpoint equation \#2: } \Pr(y \leq 2|\mathbf{x}) = F(\tau_2 - \mathbf{x}\boldsymbol{\beta})$$

$$\text{Cutpoint equation \#3: } \Pr(y \leq 3|\mathbf{x}) = F(\tau_3 - \mathbf{x}\boldsymbol{\beta})$$

The values for the cutpoints vary across the equations, whereas the values for the independent variables and slopes remain constant across the equations. The cutpoints are less meaningful using a nonlinear cumulative probability framework because they no longer represent key points in an underlying continuous distribution. Without the latent variable assumption, the cutpoints are simply constants in the binary regression models.

2.1.3 Interpreting the Results from Ordered Regression Models

We will rely on several methods of interpretation in this book, including odds/hazard ratios, average marginal effects (AMEs), and graphs based on predicted probabilities. The odds ratio is a commonly used measure for interpretation, but it is only available for the logit version of the ordered regression models. Using the logit link, the parallel cumulative model is a model of the log of the cumulative odds, which is the log of the odds that $y \leq m$ versus $y > m$:

$$\ln\left[\frac{\Pr(y \leq m|\mathbf{x})}{\Pr(y > m|\mathbf{x})}\right] = \tau_m - \mathbf{x}\boldsymbol{\beta} \qquad (1 \leq m < M) \qquad (2.11)$$

Taking the exponent of both sides of Equation 2.11 results in the equation for the cumulative odds:

$$\frac{\Pr(y \leq m|\mathbf{x})}{\Pr(y > m|\mathbf{x})} = \exp(\tau_m - \mathbf{x}\boldsymbol{\beta}) \qquad (1 \leq m < M) \qquad (2.12)$$

An odds ratio is a ratio of the odds for two different values of x_k. For a model with three independent variables, the odds ratio for $x_{1a} = x_1$ and $x_{1b} = x_1 + \kappa$ is

$$\frac{\exp\left[\tau_m - (\beta_1\{x_1 + \kappa\} + \beta_2 x_2 + \beta_3 x_3)\right]}{\exp\left[\tau_m - (\beta_1 x_1 + \beta_2 x_2 + \beta_3 x_3)\right]} = \frac{\exp\left[\tau_m - (\beta_1 x_1 + \beta_1 \kappa + \beta_2 x_2 + \beta_3 x_3)\right]}{\exp\left[\tau_m - (\beta_1 x_1 + \beta_2 x_2 + \beta_3 x_3)\right]}$$

where κ is the change in x_1. This simplifies to

$$\frac{\exp(\tau_m)\exp(-\beta_1 x_1)\exp(-\beta_1 \kappa)\exp(-\beta_2 x_2)\exp(-\beta_3 x_3)}{\exp(\tau_m)\exp(-\beta_1 x_1)\exp(-\beta_2 x_2)\exp(-\beta_3 x_3)} = \exp(-\beta_1 \kappa)$$

The un-standardized odds ratio is frequently used, which assumes that $\kappa = 1$:

$$\text{un-standardized odds ratio} = \exp(-\beta) \qquad (2.13)$$

An un-standardized odds ratio represents the factor change in the odds associated with a one-unit increase in x_k. The negative sign reflects the decision to parameterize the model as $\tau_m - \mathbf{x}\boldsymbol{\beta}$ rather than $\tau_m + \mathbf{x}\boldsymbol{\beta}$. This is the most appropriate odds ratio to use for binary independent variables because the only possible increase is from 0 to 1. However, for continuous independent variables, a one-unit increase may not be very meaningful, particularly if the standard deviation of x, S_x, is large. Therefore, we will also focus on the x-standardized odds ratio, which assumes that $\kappa = S_x$:

$$\text{x-standardized odds ratio} = \exp(-\beta S_x) \qquad (2.14)$$

An x-standardized odds ratio represents the factor change in the odds associated with a standard deviation increase in x_k. Another advantage of the x-standardized odds ratio is that it allows one to directly compare the relative magnitudes of odds ratios for continuous variables in the same model (see Long 1997, p. 80).

The interpretation of odds ratios is more complicated in the parallel cumulative logit model than in the binary logit model because a single odds ratio refers to several different binary outcomes. This is a result of the parallel regression assumption, which constrains the odds ratios across cutpoint equations. Although the ordinal y is recoded into $M - 1$ binary outcomes in the cutpoint equations, there is only one odds ratio for each independent variable in the parallel model. For example, if y has four categories (1–4), then the odds ratio for x_1 represents the effect of a one-unit increase in x_1 on the odds of: (a) 1 versus 2,3,4; (b) 1,2 versus 3,4; or (c) 1,2,3 versus 4. Relaxing the parallel regression assumption for x_1 would allow the odds ratios to vary by the cutpoint equation. We will consider models that relax the parallel regression assumption for one or more variables in Chapters 3 and 4.

Using the complementary log log link function, the parallel cumulative model is equivalent to the parallel continuation ratio model (Laara and Matthews 1985).[*] Additionally, if we assume that the data are generated by a continuous-time proportional hazards model, then the coefficients in these two models are equivalent to the coefficients from the proportional hazards model (Allison 1982, p. 72). The linear predictor in the continuous-time proportional hazards model is the log of the "hazard rate," which is the instantaneous probability of event occurrence at a particular point in time given that it has not already occurred (Allison 1982, p. 67). Therefore, one may interpret the exponent of the coefficient from a parallel cumulative or continuation ratio complementary log log model as a "hazard ratio" or as a factor change in the hazard rate. To illustrate, a hazard ratio of 1.5 corresponds to a 50% increase in the hazard rate. We will focus on un-standardized and x-standardized hazard ratios using the same equations for the odds ratios.

Odds and hazard ratios are useful for interpreting the results from parallel cumulative logit and complementary log log models, but there is no equivalent measure for parallel cumulative probit. Therefore, we also focus on two methods of interpretation that are common to models based on all three link functions. First, we will present graphs of the predicted probabilities for each category of y across the range of x_k. The effects of variables on the predicted probabilities in these graphs are "local" in that they require one to hold the remaining variables constant at a specific set of values. Unless otherwise noted, we will use an "at means" approach. In other words, the remaining variables are held constant at their means.

---

[*] For more details, see the Appendix.

Second, we will present AMEs in order to compare the effects of each independent variable on the predicted probabilities. For binary variables, the marginal effect is the effect of a discrete change in x_k from 0 to 1 on the predicted probability. For continuous variables, the marginal effect is the partial change in the predicted probability, which is the instantaneous rate of change or slope of the tangent line to the probability curve with respect to x_k while holding the remaining variables constant (Powers and Xie 2008, p. 60). It essentially represents a hypothetical, linear effect of a one-unit increase in x_k on the predicted probability. For continuous variables, the marginal effect will only be approximately equal to a one-unit discrete change in x_k if the change takes place in the linear portion of the curve near a probability of 0.5.

Marginal effects vary depending on the values for the remaining variables. Therefore, we rely on the AME, which averages the marginal effects across every case in the sample (Wooldridge 2010, p. 22; Long and Freese 2014, p. 243):

$$\text{AME} = \frac{1}{N} \sum_{i=1}^{N} \frac{\partial \text{Pr}\left(y_i = m \mid \mathbf{x} = \mathbf{x}_i\right)}{\partial x_k} \tag{2.15}$$

where ∂ is the partial derivative with respect to x_k. The AME is a "global" measure because it averages the effects across the entire sample (Xu and Fullerton 2014). Therefore, one does not have to decide where to hold the remaining variables constant. Local measures, such as marginal effects calculated with the remaining variables held constant "at means" (Williams 2012, p. 323), may not be as useful because they correspond to a single set of characteristics that may not be observed in the sample. AMEs average the marginal effects across every case in the sample, which one may use to draw inferences about the average effect in the population (Hanmer and Kalkan 2013). AMEs are also useful for comparing effects across models and subgroups (Mood 2010).

AMEs are easy to interpret for binary variables because they represent the average effect of a 0 to 1 increase in the sample. However, AMEs for continuous variables represent the average instantaneous rate of change, which will only correspond to the average effect of a one-unit increase in some instances. Therefore, we also report average discrete changes (ADCs) based on standard deviation increases for continuous variables. This is a global measure that calculates the effect of a standard deviation increase from the observed value in the continuous variable on the predicted probability and averages the discrete changes across every case in the sample (see Long and Freese 2014, p. 343). The equation for the ADC is

$$\text{ADC} = \frac{1}{N} \sum_{i=1}^{N} \left[\text{Pr}\left(y_i = m \mid \mathbf{x} = \mathbf{x}_i, x_k = x_{ki} + \kappa\right) - \text{Pr}\left(y_i = m \mid \mathbf{x} = \mathbf{x}_i, x_k = x_{ki}\right) \right] \tag{2.16}$$

where κ is the amount of the discrete change. In the examples, we focus on standard deviation changes for continuous variables.

Graphs of predicted probabilities and AMEs allow researchers to examine the substantive or practical significance of the effects of independent variables. Substantive significance essentially refers to a large effect size. In other words, the change in x is associated with a relatively large change in the predicted probabilities for the outcome categories. A statistically significant raw coefficient or large odds ratio does not necessarily indicate that the variable will also have a substantively significant effect. Therefore, when interpreting the results from an ordered regression model, it is preferable to focus on at least one measure based on the predicted probabilities.

2.1.4 Example: Parallel Cumulative Models of General Self-Rated Health

Health scholars often use the parallel cumulative model to examine general self-rated health within a regression framework. Self-rated health is an overall measure of perceived health from the 2010 GSS with the following ordered categories: poor, fair, good, and excellent. The latent variable assumption is appropriate for this example because it is reasonable to assume that there is an underlying continuous measure of the quality of perceived health observed at four points along the continuum. The independent variables in the parallel cumulative models of self-rated health include age, age-squared, sex, race/ethnicity, education, employment, income, and marital status. For more details regarding variable coding, see Chapter 1.

In Table 2.1, we present the results from parallel cumulative models of general self-rated health using the logit, probit, and complementary log log link functions. We present the

TABLE 2.1

Results from Parallel Cumulative Ordered Regression Models of Self-Rated Health

	(General Self-Rated Health: 1 = Poor, 2 = Fair, 3 = Good, 4 = Excellent)								
	Logit			Probit			Complementary Log Log		
	Coef.	SE	OR	Coef.	SE	Coef.	SE	HR	
Variables									
Age	−0.091***	0.020	4.511	−0.054***	0.012	−0.049***	0.013	2.241	
Age-squared[a]	0.073***	0.020	0.294	0.044***	0.011	0.038**	0.012	0.525	
Female	0.006	0.115	0.994	−0.009	0.066	−0.023	0.073	1.023	
Race/ethnicity[b]									
Black	0.101	0.176	0.904	0.044	0.101	0.101	0.113	0.904	
Latino	−0.325#	0.189	1.384	−0.195#	0.110	−0.194	0.119	1.214	
Other race/ethnicity	−0.472	0.288	1.604	−0.263	0.164	−0.222	0.179	1.248	
Education	0.106***	0.022	0.723	0.060***	0.012	0.065***	0.013	0.819	
Employed	0.640***	0.129	0.527	0.368***	0.074	0.300***	0.080	0.741	
Log income	0.311***	0.062	0.691	0.169***	0.035	0.172***	0.036	0.815	
Married	0.201	0.127	0.818	0.133#	0.073	0.087	0.080	0.917	
Cutpoints									
τ_1	−0.632	0.655		−0.493	0.381	−1.512	0.426		
τ_2	1.462	0.651		0.660	0.379	0.293	0.412		
τ_3	3.750	0.661		2.027	0.382	1.807	0.413		
Model Fit									
R_M^2		0.070			0.071		0.055		
R_O^2		0.099			0.099		0.066		
R_N^2		0.053			0.049		0.039		

Note: The results from significance tests are not reported for the cutpoints. OR = odds ratio, HR = hazard ratio. X-standardized odds and hazard ratios are reported for continuous variables (age, age-squared, education, and log income) and un-standardized odds and hazard ratios are reported for binary variables (female, race/ethnicity, employed, and married). $N = 1129$.

[a] Coefficients and standard errors are multiplied by 100 for ease of presentation.

[b] Reference = White.

#$p < 0.10$, *$p < 0.05$, **$p < 0.01$, ***$p < 0.001$.

un-standardized coefficients and standard errors for all three models. In addition, we present the un-standardized and x-standardized odds and hazard ratios for binary and continuous variables, respectively.

There are several consistent findings regarding the direction and statistical significance of the coefficients across the three models. The coefficients for age and age-squared are significant in each model. We will illustrate the significant, nonlinear effect of age in Figure 2.1. Additionally, education, employed, and log income have significant, positive effects on self-rated health. There are four health categories, which means there are three cutpoint equations based on different points in the cumulative distribution of health:

1. Poor versus fair, good, or excellent
2. Poor or fair versus good or excellent
3. Poor, fair, or good versus excellent

The coefficients and odds/hazard ratios refer to all three comparisons because of the parallel regression assumption. One could choose a specific equation for interpretation purposes, or one could refer to the effect of a variable on the odds or hazard of being in "poorer" health.

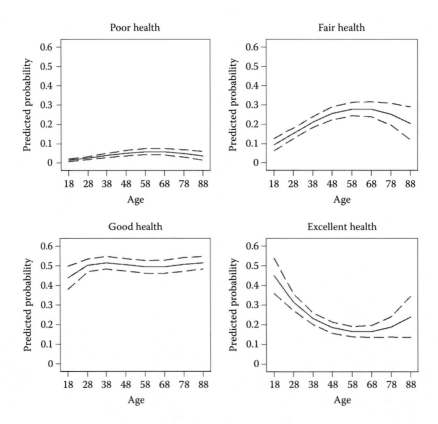

FIGURE 2.1
The effect of age on general self-rated health from a parallel cumulative logit model.

The odds ratios for education and log income are x-standardized:

Education: $\exp(-\beta S_x) = \exp(-1 \times 0.106 \times 3.066) = 0.723$

Log income: $\exp(-\beta S_x) = \exp(-1 \times 0.311 \times 1.188) = 0.691$

A standard deviation increase in education decreases the odds of poorer health by a factor of 0.72 or 28%, holding all else constant.[*] Additionally, a standard deviation increase in log income decreases the odds of poorer health by a factor of 0.69 or 31%. The x-standardized hazard ratio for log income is 0.82. Therefore, a standard deviation increase in log income decreases the hazard rate for poorer health by a factor of 0.82 or 18%.[†] Employed is a binary variable. Therefore, we present the un-standardized odds ratio:

Employed: $\exp(-\beta) = \exp(-1 \times 0.640) = 0.527$

Being employed decreases the odds of poorer health by a factor of 0.53 or 47%, and it decreases the hazard rate by a factor of 0.74 or 26%.

Despite these consistent findings, there are a few differences among the three parallel cumulative models. For example, there are marginally significant ($p < 0.10$) differences between Latinos and Whites in the logit and probit models but not in the complementary log log model, and the effect of marital status is marginally significant in the probit model and nonsignificant in the other two models. Unlike the logit and probit models, the CDF for the complementary log log model is asymmetric. Therefore, we should expect to see at least some differences in the direction and statistical significance of the coefficients across the models.

Due to the asymmetric nature of the complementary log log distribution, a reversal of the category order can affect the results beyond a reversal of the sign on the coefficients. Although the categories are ordered from 1 = poor to 4 = excellent, reversing the category order so that 1 = excellent and 4 = poor would be reasonable given that there is no natural starting point or lowest value for self-rated health. Reversing the category order does not change the absolute value of the z statistics for the logit and probit models, but it does affect the z statistics in the complementary log log model. Reversing the category order would result in marginally significant ($p < 0.10$) coefficients for Latino and other race and in a statistically significant coefficient for married ($p < 0.05$). The absolute value of the coefficient for married would increase by a factor of 2.02 (from 0.087 to −0.175). Due to the sensitivity of the complementary log log coefficients to the choice of a starting point for y, one should estimate both models (original and reversed category order) if the choice of a starting point is arbitrary.

According to the model fit statistics, this group of variables reduces the model deviance by 6%–7%, based on the R^2_M. The model fit is better according to the ordinal dispersion measure. The logit and probit models explain approximately 10% of the ordinal variation, whereas the complementary log log model only explains 7% of the ordinal variation. Additionally, the models explain 4%–5% of the nominal variation. One should expect the parallel cumulative models to explain more ordinal than nominal variation given that

---

[*] For the remaining interpretations of odds and hazard ratios in this book, we assume that the other variables in the model are held constant.

[†] Hazard ratios may be less appropriate for general self-rated health because it requires the assumption that health is the result of a sequence of irreversible stages.

they are ordinal models in the strictest sense due to the application of the parallel regression assumption.

In order to examine the substantive significance of the findings for these three parallel cumulative models, we present AMEs in Table 2.2 based on the parallel cumulative logit model. We also present ADCs in parentheses for the continuous variables, which represent the average effect of a standard deviation increase. Variables that have statistically significant coefficients in Table 2.1 also have significant AMEs for at least one health category in Table 2.2. The AMEs will give us a better sense of the substantive effect sizes. Employment status is the only variable that has a significant effect on the predicted probability for each health category. It is negatively associated with the probabilities of poor and fair health and positively associated with the probabilities of good and excellent health. On average, employment is associated with a decrease of 0.033 in the predicted probability of poor health and an increase of 0.106 in the predicted probability of excellent health. All else equal, employed individuals tend to be in better health than the unemployed. The patterns are similar for education and income in terms of the direction of the effects, but the effects on good health are not significant. A standard deviation increase in education is associated with an average increase of 0.059 in the predicted probability of excellent health. Overall, the AMEs and ADCs provide evidence that the statistically significant coefficients from the previous table are also substantively important.

Figure 2.1 presents the nonlinear effect of age on the predicted probabilities for each health category.* Graphs of predicted probabilities are particularly useful for variables

TABLE 2.2

Average Marginal Effects from a Parallel Cumulative Logit Model of Self-Rated Health

	Poor	Fair	Good	Excellent
Age	0.005***	0.012***	−0.001	−0.015***
	(0.137)	(0.174)	(−0.137)	(−0.175)
Age-squared[a]	−0.004***	−0.009***	0.001	0.012***
	(−0.039)	(−0.130)	(−0.076)	(0.245)
Female	−0.000	−0.001	0.000	0.001
Race/ethnicity[b]				
Black	−0.005	−0.013	0.001	0.017
Latino	0.018	0.043#	−0.009	−0.052#
Other race/ethnicity	0.029	0.062	−0.018	−0.073#
Education	−0.005***	−0.014***	0.001	0.018***
	(−0.014)	(−0.040)	(−0.004)	(0.059)
Employed	−0.033***	−0.088***	0.015*	0.106***
Log income	−0.016***	−0.040***	0.003	0.053***
	(−0.016)	(−0.046)	(−0.005)	(0.067)
Married	−0.010	−0.026	0.002	0.034

Note: The numbers in parentheses are average standard deviation discrete changes for continuous variables.

a The coefficient is multiplied by 100 for ease of presentation.

b Reference = White.

#$p < 0.10$, *$p < 0.05$, **$p < 0.01$, ***$p < 0.001$.

---

* The predicted probabilities and 95% confidence intervals in this chapter are based on the models using the logit link with the remaining variables held constant at their means.

measured in the quadratic form, $X + X^2$, because they allow one to present the complete pattern of the nonlinear relationship. Self-reported health declines from ages 18 to 62 and then begins to improve. For example, the predicted probability of excellent health decreases from 0.45 at age 18 to 0.16 at age 62 and then increases to 0.24 by age 88. Additionally, the predicted probability of fair health increases from 0.10 at age 18 to 0.28 at age 62 and then decreases to 0.21 by age 88. Age has a much smaller effect on the predicted probability of poor health, which may be due in part to the relatively small number of respondents in this category (5.58%). In addition, older respondents are more likely to describe their health as good, but this change takes place primarily between the ages of 18 and 38. The predicted probabilities for this category remain relatively constant from ages 38 to 88.

The confidence intervals vary in size across the age range for most health categories. The level of precision for the predicted probabilities tends to decrease as age increases for poor and fair health, whereas it is highest for excellent health in the middle of the age distribution. However, the level of precision is relatively constant across the age range for good health. The confidence intervals are widest for younger and older respondents with excellent health. It would not be possible to uncover these nuanced patterns using raw coefficients, odds/hazard ratios, or AMEs, which highlights the utility of graphs based on predicted probabilities and confidence intervals for key continuous variables.

2.2 Parallel Continuation Ratio Model

The parallel continuation ratio model is an alternative ordered regression model that focuses on the conditional probability rather than the cumulative probability (Fienberg 1980). The conditional probability (i.e., $\Pr[y = m \mid y \geq m, \mathbf{x}]$) is the probability for a particular category given that one has progressed to that stage. Therefore, we adopt Fullerton's (2009) terminology and refer to the continuation ratio models as "stage" models. The parallel continuation ratio model treats the ordinal outcome as a series of stages with effects that are constrained to remain constant across stages. In other words, the parallel regression assumption is imposed for every variable. One important difference between cumulative and stage models is that in stage models higher categories of y may only be reached after successfully proceeding through the lower categories. Scholars do not typically use stage models to analyze variables such as self-rated health because there is no natural starting point, and one does not need to experience poor or fair health before good or excellent health. However, for other variables, such as educational attainment, stage models are appropriate because one must earn a high school degree or its equivalent before earning a post-secondary degree.

As we discussed in Chapter 1, the original, ordinal dependent variable, y, is recoded into a set of binary outcomes corresponding to each stage. However, unlike the parallel cumulative model, the number of categories included in each binary outcome gets progressively smaller as one proceeds through the stages. For example, if there are four categories (1–4), then y is recoded into binary outcomes based on the following comparisons: 1 versus 2–4; 2 versus 3,4; and 3 versus 4. One may use a latent variable or nonlinear conditional probability framework to introduce the parallel continuation ratio model. As we will see in the next two sections, the equation for the conditional probability is the same using either framework. We will present the model using both frameworks and then consider an example from the field of educational research.

2.2.1 A Latent Variable Model

Tutz (1991) presents the parallel continuation ratio model within a latent variable framework, which is very similar to the latent variable framework for the parallel cumulative model.[*] However, there is one important difference. There are multiple latent variables in the parallel continuation ratio model. For $y = 1, \ldots, M$, there are $M - 1$ latent variables. Each latent variable, y_m^*, corresponds to a particular cutpoint equation or stage. The structural model for the set of latent continuous outcomes is (Agresti 2010, p. 97)

$$y_m^* = \mathbf{x}\boldsymbol{\beta} + e_m \tag{2.17}$$

where $\mathbf{x}$ is a vector of independent variables, $\boldsymbol{\beta}$ is a vector of coefficients, and e_m are the error terms for the m stages. The latent continuous outcomes are linked to the observed, ordinal outcome, y, through the following measurement equation (Tutz 1991, p. 278):

$$y = m, \quad \text{given } y \geq m, \ \text{if } y_m^* \leq \tau_m \tag{2.18}$$

The transition from one stage to the next occurs if the latent variable is above the threshold or cutpoint. The process operates as a sequence of binary decisions, but the sequence of events for the earlier transitions is not observable. The only value that is observed is the category for y.

Based on y_m^*, the equation for the conditional probability is

$$\Pr\left(y = m \mid y \geq m, \mathbf{x}\right) = \Pr\left(y_m^* \leq \tau_m \mid y \geq m, \mathbf{x}\right) \tag{2.19}$$

Substituting $\mathbf{x}\boldsymbol{\beta} + e_m$ for y_m^*,

$$\Pr(y = m \mid y \geq m, \mathbf{x}) = \Pr(\mathbf{x}\boldsymbol{\beta} + e_m \leq \tau_m \mid y \geq m, \mathbf{x}) \tag{2.20}$$

Then, subtracting $\mathbf{x}\boldsymbol{\beta}$ from both sides of the inequality,

$$\Pr(y = m \mid y \geq m, \mathbf{x}) = \Pr(e_m \leq \tau_m - \mathbf{x}\boldsymbol{\beta} \mid y \geq m, \mathbf{x}) \tag{2.21}$$

We will consider three versions of the parallel continuation ratio model based on different assumptions about the error distribution. Regardless, we may express the conditional probability as a function of τ_m and $\mathbf{x}\boldsymbol{\beta}$:

$$\Pr\left(y = m \mid y \geq m, \mathbf{x}\right) = \Pr\left(y_m^* \leq \tau_m \mid y \geq m, \mathbf{x}\right) = F\left(\tau_m - \mathbf{x}\boldsymbol{\beta}\right) \tag{2.22}$$

The probability for a particular category, m, is the product of the probability that $y = m$ for the current stage and the probability that $y > m$ for the earlier stages:

$$\Pr\left(y = m \mid \mathbf{x}\right) = F\left(\tau_m - \mathbf{x}\boldsymbol{\beta}\right) \prod_{j=1}^{m-1} \left[1 - F\left(\tau_j - \mathbf{x}\boldsymbol{\beta}\right)\right] \tag{2.23}$$

---

[*] Tutz (1991, 2012) refers to the parallel continuation ratio model as the "simple sequential model."

The parallel continuation ratio logit model assumes a logistic distribution, whereas the parallel continuation ratio probit and parallel continuation ratio complementary log log models assume normal and extreme value distributions, respectively. The cutpoints do not have to be ordered given that the sequential process is a series of binary outcomes conditional on making the earlier transitions. Additionally, the identifying assumption regarding β_0 and the cutpoints from the cumulative model is also necessary in the continuation ratio model. Statistical software programs such as Stata typically parameterize the model based on the assumption that $\beta_0 = 0$. One additional assumption that is necessary in the parallel continuation ratio model is that the random components in each cutpoint equation or stage must be independent (Maddala 1983, p. 51). In other words, we must assume that there is no correlation among the error terms in the different cutpoint equations.

2.2.2 A Nonlinear Conditional Probability Model

The latent variable assumption provides a convenient interpretation for the parallel continuation ratio model, but one may also view the model within the nonlinear probability framework we introduced earlier. The probability of interest in stage models is the conditional probability (i.e., $\Pr[y = m | y \geq m, \mathbf{x}]$), which is the probability that $y = m$ given that the respondent has progressed to that stage of the model. The conditional probability is nonlinearly related to the set of independent variables through the following equation:

$$\Pr(y = m \mid y \geq m, \mathbf{x}) = F(\tau_m - \mathbf{x}\boldsymbol{\beta}) \tag{2.24}$$

where F typically represents the logit, probit, or complementary log log CDF.

The original, ordinal dependent variable, y, is recoded into a set of binary outcomes corresponding to each stage or cutpoint equation. If $M = 4$, then the parallel continuation ratio model estimates the following three binary cutpoint equations simultaneously:

Cutpoint equation #1: $\Pr(y = 1 \mid y \geq 1, \mathbf{x}) = F(\tau_1 - \mathbf{x}\boldsymbol{\beta})$
Cutpoint equation #2: $\Pr(y = 2 \mid y \geq 2, \mathbf{x}) = F(\tau_2 - \mathbf{x}\boldsymbol{\beta})$
Cutpoint equation #3: $\Pr(y = 3 \mid y \geq 3, \mathbf{x}) = F(\tau_3 - \mathbf{x}\boldsymbol{\beta})$

The values for the cutpoints vary across the equations, whereas the values for the independent variables and slopes remain constant across the equations. The cutpoints are less meaningful using a nonlinear conditional probability framework because they no longer represent key points in the underlying continuous distributions for the latent variables. Without the latent variable assumption, the cutpoints are simply constants in the binary regression models. However, the methods of interpretation that we used in the previous section are not affected by the motivation for the model.

One important difference among the cutpoint equations in the parallel continuation ratio and the parallel cumulative models is the progressively smaller sample sizes in Cutpoint Equations 2 and 3 (assuming $M = 4$). In order for a case to be in the sample for Cutpoint Equation 2, the respondent must make the transition from $y = 1$ to $y = 2, 3$, or 4 in the first cutpoint equation. Using event history terminology, the respondent must "survive" to the second stage or cutpoint equation in order to be "at risk" of experiencing the event in that stage. In fact, the parallel continuation ratio model is equivalent to an event history model, the discrete-time proportional hazards model, using the complementary log log link function (Allison 1982).

2.2.3 Potential Sample Selection Bias and Scaling Effects in Stage Models

The key assumption in all stage models is that the errors are independent across stages or cutpoint equations (Maddala 1983, p. 51). For a three-category outcome with two equations, 1 versus 2–3 and 2 versus 3, we must assume that there is no correlation between the errors in the first and second equations. In other words, the smaller sample in the second equation is a random subset of the larger sample in the first equation. Unfortunately, this may not be a realistic assumption for some ordinal outcomes, such as educational attainment. If the subsamples in later stages are systematically different with respect to unobserved or omit-ted variables, then the coefficients in later stages may be biased. For instance, the finding of weakening family background effects on educational attainment in later stages is due in part to this type of sample selection bias (Cameron and Heckman 1998; Mare 2011). The group that successfully makes several transitions to the later educational stages is much more homogeneous than the initial group of respondents. One should be aware of the potential bias due to selection effects and consider conducting a sensitivity analysis to determine the severity of the problem (Buis 2011), or one should use a method such as probit with sample selection (Heckman 1979; Holm and Jæger 2011) that corrects for selection bias. However, the issue of sample selection bias is less important if one views the continuation ratio model as a descriptive rather than causal or predictive model (Mare 2011; Xie 2011).

Coefficients may also vary across cutpoint equations in stage models due to heterosce-dasticity of the conditional variances of the latent dependent variables (Cameron and Heckman 1998; Mare 2006). Ordered regression models, like binary regression models, are nonlinear probability models that require assumptions regarding the distribution of errors in order to identify the model. The coefficients are only identified up to a scale because we must assume a particular variance for the errors, such as $\frac{\pi^2}{3}$ (logit) or 1 (probit). The model ensures that the residual variance will equal a fixed value by using a scale factor, σ, which adjusts the true variance so that it equals $\frac{\pi^2}{3}$ in the logit model or 1 in the probit model.

The coefficients in the model are also divided by this scale factor. The scaled coefficient for a variable, X_j, is $\beta_j = \alpha_j / \sigma$, where α_j is the "true effect" and σ is the scale factor (Allison 1999, p. 189). We cannot directly observe α_j or σ. Instead, in nonlinear probability models we esti-mate the ratio of the true effect and scale factor, β_j. Therefore, one should not compare raw coefficients or odds ratios from nonlinear probability models such as logit or probit across models or samples due to these "scaling effects" (Allison 1999; Williams 2009; Mood 2010). The scaling issue applies to all ordered regression models. Measures based on predicted probabilities, such as AMEs, are not affected by differences in scaling based on residual variance and are therefore more appropriate for comparisons across cutpoint equations. In Chapter 6, we will introduce the heterogeneous choice model, which allows one to simul-taneously model the ordinal outcome and the residual variance (Williams 2009).

In stage models, the sample size decreases across stages because progressively fewer respondents continue to make each subsequent transition. Therefore, we should not focus on comparisons of raw coefficients across cutpoint equations because differences between coefficients could be due to differences in true effects, residual variances, or a combina-tion of both. This is more important in partial and nonparallel models, which allow one or more variables to have coefficients that vary across equations. We may view coefficient variation across equations as evidence of either variation in the effect of the variable or as differences in residual variation across equations (Mare 2006). We will return to this point in our discussion of partial and nonparallel models in Chapters 3 and 4.

2.2.4 Example: Parallel Continuation Ratio Models of Educational Attainment

Educational scholars have used continuation ratio models in order to study the sequential process of educational attainment (e.g., Mare 1980; Hauser and Andrew 2006). Educational attainment is an indicator of the highest degree achieved from the 2010 GSS with the following ordered categories: less than high school, high school or equivalent, associate or junior college, college, and postgraduate. Although earning each degree may be considered a discrete event, it is plausible that there is an underlying propensity to pursue post-secondary education. Therefore, one could also use the parallel cumulative model to examine educational attainment, but the cumulative model does not assume a sequential process. For a discussion of the relationship between odds ratios in cumulative and stage models, see Breen et al. (2009, pp. 1498–1503). The independent variables in the parallel continuation ratio models of educational attainment are age, sex, race/ethnicity, mother's education, and father's education. For more details regarding variable coding, see Chapter 1.

In Table 2.3, we present the results from parallel continuation ratio models of educational attainment using the logit, probit, and complementary log log link functions.

TABLE 2.3

Results from Parallel Continuation Ratio Ordered Regression Models of Educational Attainment

	(Educational Attainment: 1 = Less than High School, 2 = High School, 3 = Associate, 4 = College, 5 = Postgraduate)							
	Logit			**Probit**		**Complementary Log Log**		
	Coef.	**SE**	**OR**	**Coef.**	**SE**	**Coef.**	**SE**	**HR**
Variables								
Age	0.012***	0.003	0.821	0.007***	0.002	0.008***	0.002	0.873
Female	0.094	0.087	0.910	0.052	0.051	0.054	0.063	0.947
Race/ethnicity[a]								
Black	−0.212	0.153	1.236	−0.121	0.087	−0.052	0.108	1.054
Latino	−0.469**	0.170	1.598	−0.267**	0.099	−0.321**	0.117	1.378
Other race/ethnicity	0.812***	0.233	0.444	0.451***	0.134	0.583**	0.183	0.558
Parents' education								
Mother's education	0.121***	0.018	0.635	0.069***	0.010	0.069***	0.012	0.770
Father's education	0.087***	0.015	0.692	0.050***	0.009	0.052***	0.010	0.803
Cutpoints								
τ_1	0.529	0.290		0.279	0.168	−0.524	0.214	
τ_2	3.123	0.302		1.783	0.173	1.488	0.204	
τ_3	1.587	0.316		0.878	0.181	0.235	0.227	
τ_4	4.046	0.329		2.325	0.187	2.020	0.214	
Model Fit								
R^2_M	0.081			0.083		0.063		
R^2_O	0.132			0.128		0.089		
R^2_N	0.064			0.060		0.044		

Note: The results from significance tests are not reported for the cutpoints. OR = odds ratio, HR = hazard ratio. X-standardized odds and hazard ratios are reported for continuous variables (age and parents' education) and un-standardized odds and hazard ratios are reported for binary variables (female and race/ethnicity). N = 1286.

[a] Reference = White.

$^\#p < 0.10$, $^*p < 0.05$, $^{**}p < 0.01$, $^{***}p < 0.001$.

We present the un-standardized coefficients and standard errors for all three models. In addition, we present the un-standardized and x-standardized odds and hazard ratios for binary and continuous variables, respectively.

The direction and statistical significance of the effects of the independent variables are generally consistent across the three models. Age has a consistently positive effect on educational attainment. Older adults tend to have higher levels of education, which means that age actually reduces the odds of having a particular level of education compared to a higher level. A standard deviation in increase in age reduces the odds of a given educational level (vs. higher levels) by a factor of 0.82 or 18%. There are four cutpoint equations, and the parallel regression assumption is retained. Therefore, the odds and hazard ratios correspond to each of the following comparisons:

1. Less than high school versus high school or more
2. High school versus associate or more
3. Associate versus college or more
4. College versus postgraduate

Sex differences in educational attainment are consistently nonsignificant, but there are significant racial and ethnic differences. Latinos have a significantly lower level of educational attainment than Whites, and individuals that identify with a race other than White or Black have a significantly higher level of educational attainment than Whites. If we focus on the second comparison to simplify the interpretation, then we can interpret the un-standardized odds ratios for Latino as follows: Identifying as Latino (vs. White) is associated with an increase in the odds of having a high school degree compared to an associate's degree or more by a factor of 1.60 or 60%.

Parents' education also has a significant, positive effect on educational attainment. Focusing on the fourth comparison, we can interpret the x-standardized odds ratios for mother's education as follows: A standard deviation increase in mother's education is associated with a decrease in the odds of having a college degree compared to a postgraduate degree by a factor of 0.64 or 36%. Additionally, a standard deviation increase in mother's education is associated with a decrease in the hazard rate for a college degree by a factor of 0.77 or 23%. Father's education also appears to have a substantial effect on educational attainment. Although the effect of mother's education is slightly stronger, the difference in the magnitudes of these two coefficients is not statistically significant.* Finally, although the results for the complementary log log model could change by reversing the category order, it would not make sense to reverse the order given that less than high school is the natural starting point for this set of categories.

According to the model fit statistics, the sociodemographic and family background variables reduce the model deviance by 6%–8%, and they explain a fair amount of the dispersion in educational attainment. The logit and probit models explain approximately 13% of the ordinal dispersion and 6% of the nominal dispersion in education. We should expect the pseudo-R^2 to be higher for ordinal than nominal dispersion given that the models are ordinal in the strictest sense. All three parallel continuation ratio models explain twice as much ordinal dispersion as nominal dispersion. Finally, the pseudo-R^2 values are consistently lower for the complementary log log model, but this link function is more meaningful for stage models given its connection to event history models (Allison 1982).

---

* The nonsignificant difference is based on a Wald test for the equality of coefficients.

We present conditional AMEs for the parallel continuation ratio logit model in Table 2.4 in order to examine the substantive significance of the effects of key independent variables. The AMEs are conditional in that they are based on predicted probabilities from a particular cutpoint equation. For example, the conditional AMEs for the fourth category, college, are based on Cutpoint Equation 4 (college vs. postgraduate). We also present conditional ADCs for continuous variables in parentheses.

Although the AME for age is statistically significant, its substantive effect size is more modest than most of the remaining variables. For example, a standard deviation increase in age is associated with an average decrease of 0.038 in the conditional predicted probability of a college degree, which is weaker than the average effects of both indicators of parents' education. There are also significant racial differences in educational attainment, and the difference between Whites and respondents identifying with other races is particularly important.

Mother's education is one of the strongest predictors of educational attainment for the models. Father's education has similar but slightly weaker effects. Figure 2.2 presents a graph with the conditional predicted probabilities and 95% confidence intervals across the range of mother's education. The influence of mother's education is most noticeable for the conditional predicted probabilities for high school and college degrees. For respondents with a college degree or more, the predicted probability of a college degree decreases from 0.92 to 0.50 as mother's education increases from 0 to 20. The conditional predicted probability of having a high school degree also substantially decreases across the range of mother's education (from 0.81 to 0.28). The changes in the conditional predicted probabilities are smaller for less than high school and an associate's degree, which is partly due to the floor effect as the probabilities approach 0. The negative effects of mother's education on the conditional predicted probabilities indicate that respondents with highly educated mothers are more likely to make the transitions to higher educational stages.

TABLE 2.4

Conditional Average Marginal Effects from a Parallel Continuation Ratio Logit Model of Educational Attainment

	Less than High School	High School	Associate	College
Age	−0.001***	−0.003***	−0.002***	−0.002***
	(−0.015)	(−0.043)	(−0.028)	(−0.038)
Female	−0.007	−0.021	−0.014	−0.018
Race/ethnicity[a]				
Black	0.018	0.047	0.033	0.039
Latino	0.042*	0.104**	0.077*	0.082**
Other race/ethnicity	−0.049***	−0.174***	−0.099***	−0.169***
Parents' education				
Mother's education	−0.010***	−0.027***	−0.018***	−0.023***
	(−0.031)	(−0.099)	(−0.060)	(−0.091)
Father's education	−0.007***	−0.019***	−0.013***	−0.016***
	(−0.026)	(−0.081)	(−0.050)	(−0.073)

Note: The numbers in parentheses are average standard deviation discrete changes for continuous variables.

[a] Reference = White.

[#]$p < 0.10$, *$p < 0.05$, **$p < 0.01$, ***$p < 0.001$.

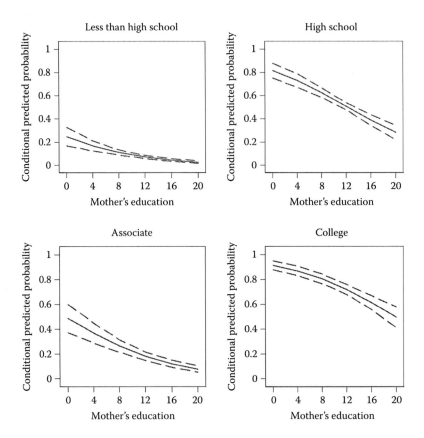

FIGURE 2.2
The conditional effect of mother's education on educational attainment from a parallel continuation ratio logit model.

There are two distinct patterns in the confidence intervals. For less than high school and an associate's degree, the precision of the estimates increases across the range of mother's education. However, for high school and college, the precision is greatest in the middle of the distribution of mother's education. The precision is lowest for respondents with a high school or associate's degree at low levels of mother's education. On balance, these results suggest that mother's education has a very strong effect on educational attainment. All else equal, respondents with highly educated mothers are much more likely to have higher levels of educational attainment than other respondents, although the precision of these estimates does vary by the levels of schooling and mother's education.

Earlier we discussed the potential for biased coefficients in stage models due to the selection process leading to progressively smaller samples in later stages and to violations of the assumption of independent errors across stages. We compared the results from the parallel continuation ratio logit model in this example to a parallel cumulative logit model and found substantively very similar results. This gives us more confidence in the results from the parallel continuation ratio models. In general, it may be useful to compare the results from an ordered regression model to similar models from other approaches. For example, one could compare the results for all three parallel models (cumulative, continuation ratio, and adjacent category). In our experience, the decision to apply the parallel regression assumption to all, some, or none of the independent variables has a larger impact on the substantive findings than the decision to use a cumulative, stage, or adjacent model. If the three parallel models produce very similar results, then factors other than model fit should determine the choice of a particular approach or class of models, including the substantive importance of the outcome categories and whether or not the outcome is the result of a sequential process.

2.3 Parallel Adjacent Category Model

The parallel adjacent category model is an ordered regression model that focuses on the adjacent probability (Goodman 1983), which is the probability for a particular category compared to the next highest category (i.e., $\Pr[y = m \mid y = m \text{ or } y = m + 1, \mathbf{x}]$). It is similar to the parallel continuation ratio model in that it is a nonlinear conditional probability model, but the focus is only on adjacent categories: m and $m + 1$. Additionally, the choice of a starting point does not affect the relative size or significance of the coefficients. Reversing the category order only changes the sign on the coefficients, which is also true for cumulative models. Adjacent category models are useful for different types of ordinal outcomes, including those that are also typically analyzed using stage models. For example, a parallel adjacent category model of educational attainment would focus on the probability of each educational level compared to the next highest level, which is a meaningful comparison. Adjacent category models are also well suited for ordinal outcomes with substantively meaningful categories given that the focus in each cutpoint equation is on the comparison of individual categories rather than on different points in the cumulative distribution.

As we discussed in Chapter 1, the original, ordinal dependent variable, y, is recoded into a set of binary outcomes corresponding to each set of adjacent categories. The sample size will vary across the binary equations unless the observations are evenly distributed across categories of y. Additionally, the focus on adjacent comparisons means that no observation is included in more than two binary cutpoint equations. To illustrate, if there are four categories (1–4), then y is recoded into binary outcomes based on the following comparisons: 1 versus 2, 2 versus 3, and 3 versus 4.

The parallel adjacent category model is presented as a series of binary regression models for the adjacent comparisons without any reference to a latent continuous variable.[*] As a result, the discrete categories are assumed to be meaningful and not simply observed points along an underlying continuous distribution. If the individual categories are of little substantive interest and the sequential logic of stage models is not appropriate, then the parallel cumulative model may be the most appropriate ordinal model to use. In the next section, we present the parallel adjacent category model as a nonlinear adjacent probability model and then consider an example from public opinion research.

---

[*] However, it is possible to use a latent variable motivation for the adjacent category model given that it is a constrained form of multinomial logit, which McFadden (1973) presented within a random utility framework. We limit our presentation to the nonlinear probability model framework, which is how the model is interpreted in the literature (e.g., Goodman 1983; Clogg and Shihadeh 1994; Agresti 2010).

2.3.1 A Nonlinear Adjacent Probability Model

The adjacent probability is nonlinearly related to the independent variables through the following equation (Agresti 2010, p. 88):

$$\Pr(y = m \mid y = m \text{ or } y = m + 1, \mathbf{x}) = F(\tau_m - \mathbf{x}\boldsymbol{\beta}) \qquad (2.25)$$

where F typically represents the logit or probit CDF.

The original, ordinal dependent variable, y, is recoded into a set of binary outcomes corresponding to each cutpoint equation. If $M = 4$, then the parallel adjacent category model estimates the following three binary cutpoint equations simultaneously:

Cutpoint equation #1: $\Pr(y = 1 \mid y = 1 \text{ or } y = 2, \mathbf{x}) = F(\tau_1 - \mathbf{x}\boldsymbol{\beta})$

Cutpoint equation #2: $\Pr(y = 2 \mid y = 2 \text{ or } y = 3, \mathbf{x}) = F(\tau_2 - \mathbf{x}\boldsymbol{\beta})$

Cutpoint equation #3: $\Pr(y = 3 \mid y = 3 \text{ or } y = 4, \mathbf{x}) = F(\tau_3 - \mathbf{x}\boldsymbol{\beta})$

The values for the cutpoints vary across the equations, whereas the values for the independent variables and slopes remain constant across the equations.

The parallel adjacent category logit model is a constrained version of the well-known multinomial logit model (Long 1997). Multinomial logit is typically presented as a baseline logit model. For an outcome with four categories (1–4), multinomial logit simultaneously estimates three binary logit equations with a common baseline category (e.g., $y = 4$):

$$\text{Log}\left(\frac{\Pr[y = 1 \mid \mathbf{x}]}{\Pr[y = 4 \mid \mathbf{x}]}\right) = \alpha_1 + \mathbf{x}\boldsymbol{\beta}_1$$

$$\text{Log}\left(\frac{\Pr[y = 2 \mid \mathbf{x}]}{\Pr[y = 4 \mid \mathbf{x}]}\right) = \alpha_2 + \mathbf{x}\boldsymbol{\beta}_2$$

$$\text{Log}\left(\frac{\Pr[y = 3 \mid \mathbf{x}]}{\Pr[y = 4 \mid \mathbf{x}]}\right) = \alpha_3 + \mathbf{x}\boldsymbol{\beta}_3$$

Using the properties of natural logs, we may derive the other two adjacent comparisons:

$$\text{Log}\left(\frac{\Pr[y=1|\mathbf{x}]}{\Pr[y=4|\mathbf{x}]}\right) - \text{Log}\left(\frac{\Pr[y=2|\mathbf{x}]}{\Pr[y=4|\mathbf{x}]}\right) = \text{Log}\left(\frac{\Pr[y=1|\mathbf{x}]}{\Pr[y=4|\mathbf{x}]} \div \frac{\Pr[y=2|\mathbf{x}]}{\Pr[y=4|\mathbf{x}]}\right) = \text{Log}\left(\frac{\Pr[y=1|\mathbf{x}]}{\Pr[y=2|\mathbf{x}]}\right)$$

$$\text{Log}\left(\frac{\Pr[y=2|\mathbf{x}]}{\Pr[y=4|\mathbf{x}]}\right) - \text{Log}\left(\frac{\Pr[y=3|\mathbf{x}]}{\Pr[y=4|\mathbf{x}]}\right) = \text{Log}\left(\frac{\Pr[y=2|\mathbf{x}]}{\Pr[y=4|\mathbf{x}]} \div \frac{\Pr[y=3|\mathbf{x}]}{\Pr[y=4|\mathbf{x}]}\right) = \text{Log}\left(\frac{\Pr[y=2|\mathbf{x}]}{\Pr[y=3|\mathbf{x}]}\right)$$

Therefore, the effect of x_k on the log odds of 1 versus 2 is $\beta_{k1} - \beta_{k2}$ and the effect of x_k on the log odds of 2 versus 3 is $\beta_{k2} - \beta_{k3}$. Using this example, the parallel adjacent category logit model is equivalent to a multinomial logit model with the following constraints:

$$\text{Constraint \#1: } \beta_{k2} - \beta_{k3} = \beta_{k3} \quad \text{or} \quad \beta_{k2} = 2\beta_{k3}$$

$$\text{Constraint \#2: } \beta_{k1} - 2\beta_{k3} = \beta_{k3} \quad \text{or} \quad \beta_{k1} = 3\beta_{k3}$$

With the application of these constraints and a slightly different parameterization of the model, $\tau_m - \mathbf{x}\boldsymbol{\beta}$ rather than $\tau_m + \mathbf{x}\boldsymbol{\beta}$, the multinomial logit model becomes equivalent to the parallel adjacent category logit model.

We also consider the probit version of the model with the independence of irrelevant alternatives (IIA) assumption, which requires that the categories are conceptually and empirically distinct (McFadden 1973; Long 1997, pp. 182–184; Cheng and Long 2007). More specifically, the IIA assumption requires that the ratio of the probabilities for categories m and $m + 1$ is unaffected if a new alternative (e.g., $m + 2$) becomes available. This assumption is necessary because the samples in the binary cutpoint equations are restricted to the cases in the two adjacent categories, m and $m + 1$. Olsen (1982, p. 521) describes the violation of the IIA assumption as a form of sample selection bias. If the ratio of probabilities for two categories is affected by the exclusion of a third category, then this suggests that the cases in the third category are systematically different and their exclusion is leading to the same type of sample selection bias that we discussed in the context of stage models.[*] The multinomial logit model relies on the IIA assumption as well. If two or more categories are similar and may be viewed as substitutes for one another, then the probability ratios are likely to be affected by the exclusion of the remaining categories. Therefore, IIA is not a realistic assumption (McFadden 1973, p. 113).[†] The IIA assumption is important to keep in mind when considering the adjacent category model. These models are only appropriate if the IIA assumption is reasonable. Unfortunately, formal tests of the IIA assumption do not have good size properties (Cheng and Long 2007). If two or more categories for an ordinal response variable are similar, such as weak and strong Democrat, then one should either combine categories or use a model that does not require the IIA assumption, such as the parallel cumulative model.

The multinomial probit model with IIA is not widely used in practice. Therefore, we focus mainly on the logit link. For the parallel adjacent category logit model, we may write the equation for the linear predictor as

$$\ln\left[\frac{\Pr(y = m \mid \mathbf{x})}{\Pr(y = m+1 \mid \mathbf{x})}\right] = \tau_m - \mathbf{x}\boldsymbol{\beta} \qquad (1 \le m < M) \tag{2.26}$$

---

[*] We thank one of the anonymous reviewers for pointing out this connection between IIA and selection bias.

[†] See the classic "red bus, blue bus" example that clearly violates the IIA assumption (Long 1997, p. 182; Powers and Xie 2008, p. 262; Train 2009, p. 46). The traditional multinomial probit model, which does not rely on the IIA assumption, requires the use of alternative specific data and is not considered in this book (see Long and Freese 2014, p. 469).

The predicted probability for a given category, m, is nonlinearly related to $\mathbf{x}$ through the following equations for categories 1 to M (Long and Cheng 2004, p. 276):

$$\Pr(y = m \mid \mathbf{x}) = \begin{cases} \dfrac{\exp\left[\displaystyle\sum_{r=m}^{M-1}(\tau_r - \mathbf{x}\boldsymbol{\beta})\right]}{1 + \displaystyle\sum_{q=1}^{M-1}\left\{\exp\left[\displaystyle\sum_{r=m}^{M-1}(\tau_r - \mathbf{x}\boldsymbol{\beta})\right]\right\}} & 1 \leq m \leq M-1 \\[4ex] \dfrac{1}{1 + \displaystyle\sum_{q=1}^{M-1}\left\{\exp\left[\displaystyle\sum_{r=m}^{M-1}(\tau_r - \mathbf{x}\boldsymbol{\beta})\right]\right\}} & m = M \end{cases}$$

(2.27)

Parallel adjacent category logit is a constrained form of multinomial logit in which the focus is on adjacent rather than baseline comparisons. Therefore, if we convert the adjacent odds into the baseline odds, then we can use the equation for predicted probabilities from multinomial logit. For example, if there are three outcome categories, then the cutpoint equations estimate the predicted log odds of 1 versus 2 and 2 versus 3. The log odds of 1 versus 3 is

$$\ln(\Omega_{1\text{vs.}3}) = \ln(\Omega_{1\text{vs.}2}) + \ln(\Omega_{2\text{vs.}3})$$

Therefore, if category 3 is the baseline category, then the predicted probabilities for categories 1–3 are

$$\Pr(y = 1 \mid \mathbf{x}) = \frac{\Omega_{1\text{vs.}3}}{\Omega_{1\text{vs.}3} + \Omega_{2\text{vs.}3} + 1} = \frac{\exp\left(\ln\left[\Omega_{1\text{vs.}2}\right] + \ln\left[\Omega_{2\text{vs.}3}\right]\right)}{\exp\left(\ln\left[\Omega_{1\text{vs.}2}\right] + \ln\left[\Omega_{2\text{vs.}3}\right]\right) + \exp\left(\ln\left[\Omega_{2\text{vs.}3}\right]\right) + 1}$$

$$\Pr(y = 2 \mid \mathbf{x}) = \frac{\Omega_{2\text{vs.}3}}{\Omega_{1\text{vs.}3} + \Omega_{2\text{vs.}3} + 1} = \frac{\exp\left(\ln\left[\Omega_{2\text{vs.}3}\right]\right)}{\exp\left(\ln\left[\Omega_{1\text{vs.}2}\right] + \ln\left[\Omega_{2\text{vs.}3}\right]\right) + \exp\left(\ln\left[\Omega_{2\text{vs.}3}\right]\right) + 1}$$

$$\Pr(y = 3 \mid \mathbf{x}) = \frac{1}{\Omega_{1\text{vs.}3} + \Omega_{2\text{vs.}3} + 1} = \frac{1}{\exp\left(\ln\left[\Omega_{1\text{vs.}2}\right] + \ln\left[\Omega_{2\text{vs.}3}\right]\right) + \exp\left(\ln\left[\Omega_{2\text{vs.}3}\right]\right) + 1}$$

The multinomial logit model may also be set up as an adjacent odds model rather than a baseline odds model. However, unlike in the parallel adjacent category logit model, the coefficients for each variable are free to vary across the equations in the multinomial logit model.

2.3.2 Example: Parallel Adjacent Category Models of Welfare Spending Attitudes

Public opinion researchers study attitudes toward government spending in several different areas. These studies often rely on a three-category ordinal measure of support or opposition to government spending. In this example, we use data from the 2012 GSS. The dependent variable is a measure of support or opposition to government spending on welfare. Respondents were asked whether spending was "too much" (1), "about right" (2), or "too little" (3). We will refer to these three categories as opposition, mixed support, and support for spending on welfare, respectively. The IIA assumption is reasonable for this example because these categories reflect three distinct views of current spending levels. The independent variables in the parallel adjacent category models are age, sex, income, education, ideology, party identification, and racial attitude. Due to the racialized nature of welfare attitudes in the United States (Gilens 1999), we limit the sample to White respondents. For more details regarding the variable coding, see Chapter 1.

In Table 2.5, we present the results from parallel adjacent category models of welfare attitudes using the logit and probit (with IIA) link functions. We present the un-standardized coefficients and standard errors for both models. In addition, we present the un-standardized and x-standardized odds ratios for binary and continuous variables, respectively. The results are similar for the two models. Welfare attitudes are most closely related to a respondent's income, party identification, and racial attitudes.

One may interpret the coefficients in terms of either of the following two cutpoint equations given the use of the parallel regression assumption:

1. Too much versus about right
2. About right versus too little

TABLE 2.5

Results from Parallel Adjacent Category Ordered Regression Models of Attitudes toward Government Spending on Welfare

| | (Attitude toward Spending on Welfare: 1 = Too Much, 2 = About Right, 3 = Too Little) | | | | |
| | Logit | | | Probit | |
	Coef.	SE	OR	Coef.	SE
Variables					
Age	0.008[#]	0.004	0.879	0.005[#]	0.003
Female	−0.012	0.151	1.012	0.005	0.114
Income	−0.239***	0.070	1.313	−0.185***	0.053
Education	−0.011	0.029	1.032	−0.014	0.022
Ideology	−0.094	0.060	1.146	−0.075[#]	0.045
Party identification[a]					
Democrat	0.936***	0.241	0.392	0.688***	0.178
Independent	0.886***	0.223	0.412	0.650***	0.164
Racial attitude	−0.244**	0.078	1.281	−0.182**	0.058
Cutpoints					
τ_1	−2.602	0.893		−2.083	0.683
τ_2	−2.015	0.864		−1.712	0.667
Model Fit					
R^2_M		0.084			0.083
R^2_O		0.120			0.116
R^2_N		0.086			0.080

Note: The results from significance tests are not reported for the cutpoints. OR = odds ratio. X-standardized odds ratios are reported for continuous variables (age, income, education, ideology, and racial attitude) and un-standardized odds ratios are reported for binary variables (female and party identification). $N = 392$.
[a] Reference = Republican.
[#]$p < 0.10$, *$p < 0.05$, **$p < 0.01$, ***$p < 0.001$.

Income is positively associated with Whites' opposition to welfare spending. A standard deviation increase in the log of family income is associated with an increase in the odds of greater opposition to welfare spending among Whites by a factor of 1.31 or 31%. A negative view of Blacks' work ethic is also related to greater opposition to welfare spending. Finally, party identification appears to be one of the most important factors shaping welfare attitudes. All else equal, the odds of greater welfare opposition are lower for Democrats than Republicans by a factor of 0.39 or 61%. The partisan gap for Independents is very similar. Overall, these results suggest that opposition to welfare spending is highest among Whites that have a high family income, identification with the Republican Party, and that hold negative racial attitudes.

The measures of model fit indicate that this group of variables reduces the model deviance by approximately 8%. Additionally, the models explain 12% of the ordinal variation and 8%–9% of the nominal variation. We expected the ordinal R^2 to be higher than the nominal R^2 given that these are ordinal models in the strictest sense.

Table 2.6 presents AMEs for each variable with ADCs for continuous variables in parentheses. Income, identification with the Republican Party, and negative racial attitudes are positively associated with the probability of welfare opposition and negatively associated with the probabilities of support and mixed support. For example, a standard deviation increase in the strength of one's negative racial attitude is associated with an average increase of 0.069 in the predicted probability of welfare opposition. There is also a substantial partisan effect on welfare attitudes. On average, identifying as a Democrat rather than a Republican is associated with an increase of 0.20 in the predicted probability of support for welfare spending.

TABLE 2.6

Average Marginal Effects from a Parallel Adjacent Category Logit Model of Attitudes toward Government Spending on Welfare

	Too Much	About Right	Too Little
Age	−0.002[#]	0.001[#]	0.001[#]
	(−0.036)	(0.010)	(0.026)
Female	0.003	−0.001	−0.002
Income	0.067***	−0.021**	−0.047***
	(0.076)	(−0.027)	(−0.048)
Education	0.003	−0.001	−0.002
	(0.009)	(−0.003)	(−0.006)
Ideology	0.027	−0.008	−0.018
	(0.038)	(−0.013)	(−0.025)
Party identification[a]			
Democrat	−0.261***	0.062***	0.199***
Independent	−0.240***	0.068***	0.172***
Racial attitude	0.069**	−0.021**	−0.048**
	(0.069)	(−0.025)	(−0.044)

Note: The numbers in parentheses are average standard deviation discrete changes for continuous variables.
[a] Reference = Republican.
[#]$p < 0.10$, *$p < 0.05$, **$p < 0.01$, ***$p < 0.001$.

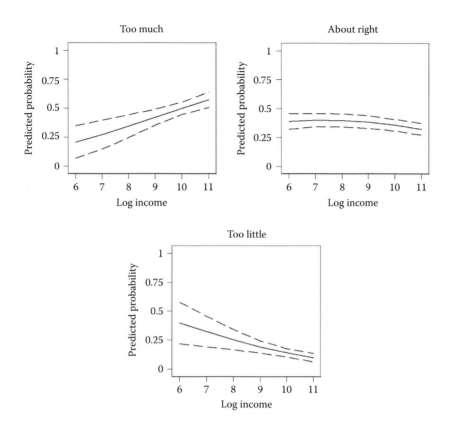

FIGURE 2.3
The effect of log income on attitudes toward government spending on welfare from a parallel adjacent category logit model.

Figure 2.3 presents the predicted probabilities and 95% confidence intervals for welfare attitudes across the range of the log of family income. As the log of income increases from a low of 6 (or $403) to a high of 11 (or $59,874), the predicted probability of opposition increases by 0.37 and the predicted probability of support decreases by 0.30 while mixed support remains relatively stable. The confidence bands indicate that the estimates are more precise at higher levels of income, although this is primarily the case for the two extreme views.

2.4 Estimation

2.4.1 Basics of Maximum Likelihood Estimation

We estimated the ordered regression models in this book using maximum likelihood estimation (MLE), which is widely used for the estimation of categorical regression models. Essentially, it is a method for finding the set of parameters that provide the "maximal" probability of observing the data, based on statistical models and assumptions.

The MLE method requires the construction of the "likelihood" or "likelihood function." Many researchers confuse likelihood with probability. Probability refers to the chance of observing an event or series of events given the population parameters, and the likelihood is proportional to the probability of the population parameters given the sample data. Because MLE typically involves maximizing a function based on population parameters, the parts of the likelihood function that do not contain these parameters (e.g., constants) are usually dropped. It is in this sense that the likelihood is proportional to the probability. MLE involves maximizing the natural log of the likelihood function because it is a monotonic increasing function and more mathematically tractable than the likelihood function. Statistical packages typically rely on numerical iterative optimization methods in order to find parameter estimates that maximize the log likelihood, such as the Newton–Raphson method (for more details, see Eliason 1993). Maximization is an iterative process that begins with a set of starting values and converges when two consecutive approximations are very close to each other according to a predetermined convergence criterion.

2.4.2 Common Problems in the use of MLE for Ordered Regression Models

Using MLE for ordered regression models is not immune to problems. Although McCullagh (1980) and Pratt (1981) show that the log-likelihood function in ordered regression models has global concavity under general conditions, the numerical estimation of these models can occasionally be problematic in practice.

When the overall sample size is small, for example, the estimation of ordered regression models could fail to reach convergence (McCullagh 1980) or to produce parameter estimates with less-than-optimal properties that are largely unknown. Long (1997, p. 54) recommended that the overall estimation sample size should be greater than 500 and that there should be roughly 10 observations per parameter. Long (1997, p. 125) also indicated that ordered regression models usually take more iterations to reach convergence than other regression models for categorical outcomes, and this is especially true when there is a small number of cases in one or more of the response categories. Fullerton and Xu (2012) also addressed the sample size issue in their discussion of the proportional odds with partial proportionality constraints model by providing a few ad hoc solutions that largely rely on the parallel regression assumption. Because there is little empirical knowledge about the possible bias or inconsistency introduced by these ad hoc solutions, one should estimate the ordered regression model using a few alternative methods and then triangulate the results.

Another possible cause of nonconvergence in the estimation of ordered regression models is (near) perfect prediction (Greene and Hensher 2010; Long and Freese 2014), or complete separation and quasi-complete separation (Albert and Anderson 1984; Allison 2008). In other words, a linear combination of independent variables can (nearly) perfectly predict some or all outcome categories. In a simple trivariate case regressing self-reported health (1 = poor, 2 = fair, 3 = good, 4 = excellent) on race (1 = White; 0 = Black) and age (in years), for example, if all Blacks reported having fair health or no Blacks reported having excellent health, then most statistical packages would usually stop the estimation process or would provide partial results because of the problem of "complete separation" in the data. Long and Freese (2014), Allison (2008), and Greene and Hensher (2010) consider possible solutions, including dropping cases that cause such problems, combining categories, or using exact inference.

Another issue that may occasionally arise is that the cutpoints become unordered during ML estimation.[*] In the parallel cumulative model, the cutpoints must be ordered (i.e., $\tau_1 <$ $\tau_2 ... < \tau_{M-1}$) in order for the predicted probabilities to be nonnegative. Greene and Hensher (2010, p. 175) propose using an exponential function to re-parameterize the cutpoints such that $\tau_m = \tau_{m-1} + \exp(\alpha_m)$. The model directly estimate α's instead of τ's. One can easily derive $\hat{\tau}$'s using the given parametric function and their standard errors using standard asymptotic techniques, such as the delta method.

2.4.3 Person-Level and Person/Threshold-Level Estimation

The estimation theory underlying ordered regression models is well established. Assuming an individual-level unit of analysis, we refer to the standard estimation approach as a "person-level" approach because each case or person contributes one observation in the sample data. Standard statistical software packages typically include options for person-level estimation of the parallel cumulative model. However, there are several ordered regression models that are not included in most statistical software packages. For example, constrained partial models require the application of constraints and estimation of additional parameters that are not typically included in the commands for ordered regression models (for an exception, see Yee 2010). Less restrictive models, such as the nonparallel cumulative models, also have limited software options. Therefore, researchers often rely on an alternative estimation approach, which we refer to as the "person/threshold-level" approach.

The person/threshold-level approach (Cole and Ananth 2001; Choi and Cole 2004; Cole et al. 2004) estimates the ordered regression model using binary regression techniques and a person-threshold sample created through data expansion. This allows the researcher to estimate several different ordered regression models using a binary regression model, which is available in any standard statistical software package. The datasets for each binary cutpoint equation are treated as separate data sources that are appended to form the person/threshold-level dataset. For an outcome with M categories, there are $M - 1$ cutpoint equations and therefore $M - 1$ datasets that one appends to form the person-threshold file. The outcome in each threshold-specific file is a binary variable, y_m, that indicates whether a particular threshold has been reached. The thresholds are $y \le m$, $y = m \mid y \ge m$, and $y = m \mid y = m$ or $y = m + 1$ for cumulative, stage, and adjacent models, respectively. For respondents with observations in multiple threshold-specific files, the values for the independent variables are the same, but the cutpoint and binary outcome values vary across threshold files.

In general, the sample size in the person-threshold file is the sum of the sample sizes from the separate cutpoint models. If we use c_m to denote the cutpoint equation and N_{c_m} to denote the sample size in each cutpoint equation, then we have the following equation for the person-threshold total sample size: $N_{p-t} = \sum_{m=1}^{M-1} N_{c_m}$. For a cumulative model with M categories and an initial sample size of N, this equation simplifies to $(M - 1)N$. For example, if $M = 4$ and $N = 1,000$, then the sample size for the person-threshold file will be $(4-1)1,000$ or 3,000. In this example, each person contributes three observations to the person-threshold file because there are three cutpoint equations.

---

[*] This problem only applies to the cumulative approach. Stage and adjacent models do not require ordered cutpoints.

However, the person-threshold sample sizes are smaller for the stage and adjacent approaches because each cutpoint equation only includes a subset of the outcome categories. For instance, if $M = 4$ and $N = 1,000$, and the cases are evenly distributed across the categories of y (i.e., 250 cases in each category), then stage approach models would only have $1,000 + 750 + 500$ or $2,250$ cases in the person-threshold file. Using the same example, the adjacent approach models would only have $500 + 500 + 500$ or $1,500$ cases in the person-threshold file. The sample sizes get progressively smaller in the cutpoint equations for stage models, but there is no consistent pattern across equations for adjacent models. The pattern for adjacent models depends on the marginal distribution of y.

In the person-threshold data, the cutpoints are binary variables corresponding to the portion of the expanded data representing a particular cutpoint equation. In order to estimate a parallel model, the binary outcome is regressed on the cutpoint indicators and the independent variables.* The effects of the independent variables and cutpoints are treated as additive effects, which means that the effect of each independent variable is constant across cutpoint equations. In other words, the parallel regression assumption is imposed for every variable in the model. The inclusion of interaction terms between the cutpoint indicators and the independent variables effectively relaxes the parallel regression assumption. Partial models include interaction terms for a subset of variables, and nonparallel models include interaction terms for every independent variable.

Figures 2.4 and 2.5 are examples of person-level and person/threshold-level datasets for a sample of five respondents. Health is the ordinal dependent variable with four categories (1–4), and the independent variables are age and female. In order to estimate a parallel cumulative model of health using a person-level estimator, one would use the person-level dataset in Figure 2.4. There are five respondents in the sample, and there is one observation for each respondent. The ordered regression model converts the dependent variable into a series of binary outcomes and then estimates the binary equations simultaneously.

The person/threshold-level estimator requires the use of data expansion to create a dataset with cutpoint-specific observations for each respondent. In this example, there are four outcome categories and five respondents. For a cumulative ordered regression model, the person/threshold-level dataset contains 15 cases because there are three cases for each of the five respondents. The three cases correspond to the observations from each cutpoint equation. The values for the independent variables remain constant within persons, but the values for the cutpoint indicators and new binary outcome vary within persons (see Figure 2.5). The person/threshold-level approach estimates the parallel ordered regression

Person	Health	Age	Female
1	2	71	1
2	3	46	1
3	1	80	1
4	3	21	0
5	4	58	0

FIGURE 2.4
A person-level dataset.

---

* See the measurement equations linking the original, ordinal outcome to the binary outcomes for each cutpoint equation in Chapter 1. Identification requires one to assume that the intercept or one of the cutpoints is equal to 0.

Person	Cutpoint 1	Cutpoint 2	Cutpoint 3	Health	Outcome	Age	Female
1	1	0	0	2	0	71	1
1	0	1	0	2	1	71	1
1	0	0	1	2	1	71	1
2	1	0	0	3	0	46	1
2	0	1	0	3	0	46	1
2	0	0	1	3	1	46	1
3	1	0	0	1	1	80	1
3	0	1	0	1	1	80	1
3	0	0	1	1	1	80	1
4	1	0	0	3	0	21	0
4	0	1	0	3	0	21	0
4	0	0	1	3	1	21	0
5	1	0	0	4	0	58	0
5	0	1	0	4	0	58	0
5	0	0	1	4	0	58	0

FIGURE 2.5
A person/threshold-level dataset.

model as a binary regression model with the new binary outcome and the cutpoint indicators included as independent variables in addition to age and female. The new binary outcome corresponds to the binary outcomes from the separate cutpoint equations. For cases from the first cutpoint equation (Cutpoint 1 = 1), outcome = 1 if health = 1. For cases from the second cutpoint equation (Cutpoint 2 = 1), outcome = 1 if health ≤ 2. Finally, for cases from the third cutpoint equation (Cutpoint 3 = 1), outcome = 1 if health ≤ 3. In order to relax the parallel regression assumption, one would include interaction terms between the cutpoint indicators and the independent variables.

The main advantage of the person/threshold-level approach is that one can estimate a wide variety of ordered regression models, including the 12 models in our typology, using any standard statistical software package. The software package must have an option for binary regression models such as logit or probit. However, this approach also has disadvantages. First, the use of data expansion creates dependence among the observations for the cumulative approach models (Cole et al. 2004; Fullerton and Xu 2012). One may estimate the cutpoint equations separately for the stage and adjacent models because they are asymptotically independent (Fienberg 1980, p. 111), and the results from separate regressions give consistent estimates of the results from models that simultaneously estimate the cutpoint equations (Begg and Gray 1984; Clogg and Shihadeh 1994, p. 146). Additionally, one may append the person-threshold files because the observations are conditionally independent (Cole and Ananth 2001, p. 1380).

However, the equations for cumulative approach models are correlated, which means that the observations within persons are dependent in the expanded person-threshold dataset. Models for clustered data are necessary in order to adequately address this problem, although this dependence does not affect the results from the nonparallel cumulative model (Cole et al. 2004, p. 173). Scholars have used several methods that correct for this error dependence, including robust standard errors that adjust for the clustering of observations (Fullerton 2009) and generalized estimating equations (GEE) (Cole et al. 2004). The main disadvantage associated with the use of GEE or robust standard errors is that they require the use of estimating equations or pseudo-likelihood rather than MLE, which is more efficient (Cole et al. 2004, p. 177; Fullerton and Xu 2012, p. 193). Overall, person/threshold-level estimators may be useful for practical considerations such as software limitations, but person-level estimators are preferred when available due to the efficiency gains associated with the simultaneous estimation of cutpoint equations and the utilization of MLE.

2.5 Conclusions

Parallel models allow researchers to analyze ordinal dependent variables within a regression framework. In this chapter, we illustrated the parallel models with three empirical examples. We analyzed self-rated health using the parallel cumulative model and found that it is associated with age and several indicators of socioeconomic status, including education, employment, and income. We used the parallel continuation ratio model to examine educational attainment given the sequential nature of the attainment process. These models revealed that parents' education is closely related to the respondent's education. Finally, we analyzed Whites' attitudes toward welfare spending using the adjacent category model and found that welfare opposition is most closely associated with income, Republican Party identification, and negative racial attitudes.

The key assumption of parallel regressions allows the models to remain relatively parsimonious, and it maintains a strict stochastic ordering. In other words, parallel ordered regression models are "ordinal" models in the strictest sense because of the constraints placed on the coefficients. They are also parsimonious compared to the partial and nonparallel models because they only require one coefficient per independent variable (in addition to the cutpoints). However, the parallel regression assumption is often violated in practice. It is possible that one or more of the variables associated with health, education, or welfare attitudes have effects that vary across outcome levels. For example, a variable such as employment, which reduces the odds of poor health, may not increase the odds of excellent health. In Chapter 3, we introduce partial ordered regression models, which relax the parallel regression assumption for a subset of independent variables. We consider unconstrained and constrained versions of the partial models and the connections between these two applications of the parallel regression assumption.

2.6 Appendix

2.6.1 Equivalence of Two Parallel Complementary Log Log Models

Laara and Matthews (1985) show that the parallel cumulative and parallel continuation ratio models are equivalent using the complementary log log link. The cutpoints are slightly different, but the coefficients for each independent variable are exactly the same. In this Appendix, we expand upon the proof found in Laara and Matthews (1985). The equations for the parallel cumulative and continuation ratio models using the complementary log log link are

$$\text{Parallel cumulative model: } \log\left(-\log\left\{1 - \Pr\left(y \le m \mid \mathbf{x}\right)\right\}\right) = \tau_m - \mathbf{x}\boldsymbol{\beta} \qquad (2.28)$$

Parallel continuation ratio: $\log\left(-\log\left\{1-\dfrac{\Pr(y=m\,|\,\mathbf{x})}{1-\Pr(y\le m-1\,|\,\mathbf{x})}\right\}\right)=\theta_m-\mathbf{x}\boldsymbol{\eta}$ (2.29)

where τ and θ are the cutpoints, $\mathbf{x}$ is the vector of independent variables, and $\boldsymbol{\beta}$ and $\boldsymbol{\eta}$ are vectors of coefficients. Based on Laara and Matthews (1985, p. 206), we can rewrite Equation 2.29 in order to show that $\boldsymbol{\beta}=\boldsymbol{\eta}$. First, we can express Equation 2.29 in terms of cumulative probabilities:

$$\log\left(-\log\left\{\frac{1-\Pr(y\le m-1\,|\,\mathbf{x})-\Pr(y=m\,|\,\mathbf{x})}{1-\Pr(y\le m-1\,|\,\mathbf{x})}\right\}\right)=\theta_m-\mathbf{x}\boldsymbol{\eta}$$

$$\log\left(-\log\left\{\frac{1-\Pr(y\le m\,|\,\mathbf{x})}{1-\Pr(y\le m-1\,|\,\mathbf{x})}\right\}\right)=\theta_m-\mathbf{x}\boldsymbol{\eta}$$

$$\log\left(\left[-\log\left\{1-\Pr(y\le m\,|\,\mathbf{x})\right\}\right]-\left[-\log\left\{1-\Pr(y\le m-1\,|\,\mathbf{x})\right\}\right]\right)=\theta_m-\mathbf{x}\boldsymbol{\eta}$$ (2.30)

If we use $\lambda_m\{x\}$ to represent $-\log\{1-\Pr(y\le m\,|\,\mathbf{x})\}$, then we can rewrite Equation 2.30 as follows:

$$\log(\lambda_m\{x\}-\lambda_{m-1}\{x\})=\theta_m-\mathbf{x}\boldsymbol{\eta}$$

$$\log(\lambda_m\{x\})-\log(\lambda_m\{x\})+\log(\lambda_m\{x\}-\lambda_{m-1}\{x\})=\theta_m-\mathbf{x}\boldsymbol{\eta}$$

$$\log(\lambda_m\{x\})+\log\left(\frac{\lambda_m\{x\}-\lambda_{m-1}\{x\}}{\lambda_m\{x\}}\right)=\theta_m-\mathbf{x}\boldsymbol{\eta}$$

Then, substituting $-\log\{1-\Pr(y\le m\,|\,\mathbf{x})\}$ for $\lambda_m\{x\}$ in the first term and $\exp(\tau_m-\mathbf{x}\boldsymbol{\beta})$ for $\lambda_m\{x\}$ in the second term:

$$\log\left(-\log\left\{1-\Pr(y\le m\,|\,\mathbf{x})\right\}\right)+\log\left(1-\frac{\exp(\tau_{m-1}-\mathbf{x}\boldsymbol{\beta})}{\exp(\tau_m-\mathbf{x}\boldsymbol{\beta})}\right)=\theta_m-\mathbf{x}\boldsymbol{\eta}$$

$$\log\left(-\log\left\{1-\Pr(y\le m\,|\,\mathbf{x})\right\}\right)+\log\left(1-\exp\left(\left[\tau_{m-1}-\mathbf{x}\boldsymbol{\beta}\right]-\left[\tau_m-\mathbf{x}\boldsymbol{\beta}\right]\right)\right)=\theta_m-\mathbf{x}\boldsymbol{\eta}$$

$$\log\left(-\log\left\{1-\Pr(y\le m\,|\,\mathbf{x})\right\}\right)+\log\left(1-\exp(\tau_{m-1}-\tau_m)\right)=\theta_m-\mathbf{x}\boldsymbol{\eta}$$

$$\log\left(-\log\left\{1-\Pr(y\le m\,|\,\mathbf{x})\right\}\right)=\theta_m-\left[\mathbf{x}\boldsymbol{\eta}+\log\left(1-\exp(\tau_{m-1}-\tau_m)\right)\right]$$

The term added to $\mathbf{x}\boldsymbol{\eta}$ is based solely on the cutpoints. Therefore, the coefficients for the independent variables will be the same in both models (i.e., $\boldsymbol{\beta}=\boldsymbol{\eta}$), but the cutpoints will be adjusted by $\log(1-\exp(\tau_{m-1}-\tau_m))$. However, if we relax the parallel assumption for one or more independent variables, then the second term will also include the independent variables and their coefficients because $\mathbf{x}\boldsymbol{\beta}_1 \neq \mathbf{x}\boldsymbol{\beta}_2$. In other words, the cumulative and continuation ratio models using the complementary log log link are only equivalent if the parallel regression assumption is imposed for every independent variable in the model. For the partial and nonparallel models, the cumulative and continuation ratio models with the complementary log log link will *not* be equivalent even if they yield similar results (Tutz 1991, p. 286).

2.6.2 Stata Codes for Parallel Ordered Logit Models

Parallel Cumulative Logit Model of Self-Rated Health

```
ologit ghealth age agesq female black hisp2 othrace educ employed lninc
   married
```

Parallel Continuation Ratio Logit Model of Educational Attainment

```
ocratio degree5 age female black hisp2 othrace maeduc paeduc
```

Parallel Adjacent Category Logit Model of Welfare Attitudes

```
constraint 1 [too_much]age      = -1*[too_little]age
constraint 2 [too_much]female   = -1*[too_little]female
constraint 3 [too_much]educ     = -1*[too_little]educ
constraint 4 [too_much]lninc    = -1*[too_little]lninc
constraint 5 [too_much]polviews = -1*[too_little]polviews
constraint 6 [too_much]democrat = -1*[too_little]democrat
constraint 7 [too_much]indep    = -1*[too_little]indep
constraint 8 [too_much]workblks = -1*[too_little]workblks
mlogit natfare2 age female educ lninc polviews democrat indep workblks,
   base(2) constraint(1/8)
```

2.6.3 R Codes for Parallel Ordered Logit Models

Parallel Cumulative Logit Model of Self-Rated Health

```
poLogit <- vglm(as.ordered(ghealth) ~ age + agesq + female
        + black + hisp2 + othrace
        + educ + employed + lninc + married,
        cumulative(link="logit",
            parallel = TRUE,
            reverse = TRUE),
        trace = TRUE, # Rank = 1,
        data = mydta)
summary(poLogit)
```

Parallel Continuation Ratio Logit Model of Educational Attainment

```
crPoLogit <-vglm(as.ordered(degree5) ~ age + female + black
            + hisp2 + othrace + maeduc + paeduc,
        cratio(link="logit",
```

```
                parallel = TRUE,
                reverse = FALSE),
        trace = TRUE, # Rank = 1,
        data = mydta)
summary(crPoLogit)
```

Parallel Adjacent Category Logit Model of Welfare Attitudes

```
acPoLogit <- vglm(as.ordered(natfare2) ~ age + female + educ
                + lninc + polviews + democrat
                + indep + workblks,
        acat(link="loge",
                parallel = TRUE,
                reverse = FALSE),
        trace = TRUE, # Rank = 1,
        data = mydta)
summary(acPoLogit)
```

3

Partial Models

Partial models are ordered regression models that relax the parallel regression assumption for one subset of variables. Relaxing the parallel assumption allows the coefficients in this group to vary across the cutpoint equations. In other words, some variables will have coefficients that change depending on the level of the dependent variable. The parallel regression assumption ensures that the model is "ordinal" in the strictest sense (see McCullagh 1980, pp. 115–116). Therefore, partial models are only strictly ordinal with respect to the set of independent variables with the parallel assumption. Ordinal patterns may emerge for the nonparallel subset, but there is no guarantee that they will. As a result, partial models are often a mix of ordinal and more complex, nonordinal relationships. Models that allow for more complex relationships may be particularly useful in the social and behavioral sciences given that many of the response variables are "assessed" ordinal variables with observed responses that are the result of the judgments of assessors (e.g., survey respondents) often using several different criteria (Anderson 1984, p. 2). For this type of ordinal variable, it is not necessarily clear whether the ordering of outcome categories is important for the relationships with the independent variables in the model (Anderson 1984, p. 1; Long 1997, p. 115). Partial models are flexible enough to allow some variables in the model to have effects that do not maintain a strict stochastic ordering.

There are three different types of partial models: cumulative, stage, and adjacent. These models differ from one another based on the probability of interest. The choice of a particular partial model should be based on factors such as the desired comparisons and the data-generating process. Cumulative models are the most widely used in the social and behavioral sciences, but stage models are useful for ordered outcomes that are the result of a sequential decision-making process, and adjacent models are useful when the specific outcome categories are substantively meaningful. For many ordinal outcomes, comparisons of adjacent categories are more meaningful and easier to interpret than different points in the cumulative distribution. This is important in partial models given that some of the variables have coefficients that vary across equations. Interpreting equation-specific results is easier if the comparisons in each cutpoint equation are substantively meaningful. As a result, adjacent category models may be particularly useful. Researchers should also consider stage and adjacent partial models given the potential for negative predicted probabilities in the partial cumulative model (see Williams 2006, 2015).

3.1 Unconstrained versus Constrained Partial Models

Peterson and Harrell (1990) introduced the partial proportional odds model as an alternative to the proportional odds model within the class of cumulative models. Additionally, they proposed unconstrained and constrained versions of the model. The presence or absence of constraints refers to the subset of variables that do not impose the parallel regression assumption. In unconstrained partial models, the coefficients for this group freely vary across cutpoint equations. However, in constrained partial models, one or more constraints are placed on this variation across equations. The unconstrained and constrained partial models are also known as vector generalized linear models and reduced rank vector generalized linear models, respectively (Yee 2015). Peterson and Harrell (1990, p. 209) suggested that one could constrain the effects in partial models so that they varied in a linear fashion across equations. Similarly, Capuano and Dawson (2013) developed a trend odds model in which the odds for a single predictor increase or decrease in a monotonic manner across equations. Alternatively, the coefficients for a group of independent variables could vary by a set of common factors that do not have to follow linear or monotonic patterns. Hauser and Andrew (2006) used the common factor approach in their partial proportionality constraint formulation of stage models, and we extended this common factor approach to cumulative and adjacent models (see Fullerton 2009; Fullerton and Xu 2012, 2015).

Constrained partial models are the most versatile ordered regression models that we present in this book. They allow the researcher to impose the parallel regression assumption for one set of variables, relax it without additional constraints for a second set, and relax it with one or more constraints for a third set of variables. The additional constraints for the third set of variables means that the model is at least slightly more parsimonious than the unconstrained partial model. The only ordered model that is more parsimonious than the constrained partial model is the parallel model, which requires the parallel assumption for every independent variable. In this chapter, we consider unconstrained and constrained partial models within all three approaches. In addition, we illustrate the different features of the partial models with empirical examples.

3.2 Partial Cumulative Models

Partial cumulative models are an extension of the parallel cumulative model that relaxes the parallel regression assumption for a subset of the independent variables. This has two important consequences for cumulative models that do not apply to stage or adjacent models. First, negative predicted probabilities are possible in partial and nonparallel cumulative models because the nonparallel cumulative probability curves must eventually intersect (McCullagh and Nelder 1989, p. 155). In order for the predicted probabilities to fall within the 0–1 range, the log odds for equation m must be greater than the log odds for the previous cutpoint equation, $m - 1$. If the lines corresponding to the cumulative probabilities do not intersect within the set of observed values for the independent variables, then this internal inconsistency in the models may not be a serious problem. In addition, Hedeker et al. (1999, pp. 63–64) note that negative predicted probabilities are only possible in partial and nonparallel cumulative models with continuous independent variables. The predicted probabilities will be nonnegative if one uses the parallel constraint for every continuous variable or recodes the continuous variables into binary variables. Alternatively, one could use a hierarchical ordered probit model, which is an internally consistent model that relies on a nonlinear specification (Greene and Hensher 2010, p. 214).

Relaxing the parallel assumption for cumulative models also means that the latent variable motivation is no longer appropriate (see Greene and Hensher 2010, pp. 190–192). However, one may interpret the results from cumulative models without reference to a latent, continuous variable (McCullagh 1980, p. 110). We present the partial models for all three approaches in this chapter as nonlinear probability models, which does not require a latent variable assumption.

3.2.1 Unconstrained Partial Cumulative Model

The unconstrained partial cumulative model is a nonlinear cumulative probability model that relaxes the parallel regression assumption for one subset of variables. The general form of the equation for the nonlinear cumulative probability model is

$$\Pr(y \leq m \mid \mathbf{x}, \boldsymbol{\omega}) = F(\tau_m - \mathbf{x}\boldsymbol{\beta} - \boldsymbol{\omega}\boldsymbol{\eta}_m) \qquad (1 \leq m < M) \tag{3.1}$$

where F typically represents the cumulative distribution function (CDF) for the logit, probit, or complementary log log model; τ_m are cutpoints; $\mathbf{x}$ and $\boldsymbol{\omega}$ are vectors of independent variables; $\boldsymbol{\beta}$ is a vector of coefficients that do not vary across cutpoint equations; and $\boldsymbol{\eta}_m$ is a vector of coefficients that freely vary across cutpoint equations.

The original, ordinal outcome is recoded into a series of $M - 1$ binary outcomes corresponding to each cutpoint equation. If $M = 4$, then the unconstrained partial cumulative model estimates the following three binary cutpoint equations simultaneously based on different points in the cumulative distribution:

Cutpoint equation #1: $\Pr(y \leq 1 \mid \mathbf{x}, \boldsymbol{\omega}) = F(\tau_1 - \mathbf{x}\boldsymbol{\beta} - \boldsymbol{\omega}\boldsymbol{\eta}_1)$

Cutpoint equation #2: $\Pr(y \leq 2 \mid \mathbf{x}, \boldsymbol{\omega}) = F(\tau_2 - \mathbf{x}\boldsymbol{\beta} - \boldsymbol{\omega}\boldsymbol{\eta}_2)$

Cutpoint equation #3: $\Pr(y \leq 3 \mid \mathbf{x}, \boldsymbol{\omega}) = F(\tau_3 - \mathbf{x}\boldsymbol{\beta} - \boldsymbol{\omega}\boldsymbol{\eta}_3)$

The values for the cutpoints (τ_m) and the coefficients for the second set of independent variables ($\boldsymbol{\eta}_m$) vary across equations; whereas the values for the independent variables ($\mathbf{x}$ and $\boldsymbol{\omega}$) and the coefficients for the first set of variables ($\boldsymbol{\beta}$) remain constant across equations. This version of the partial model is unconstrained in the sense that the values for $\boldsymbol{\eta}_m$ do not vary across equations according to a prespecified pattern, such as a linear increase or decrease. Additionally, the variation in the coefficients for one variable in $\boldsymbol{\omega}$ is not dependent on the variation for other variables in the subset.

3.2.2 Constrained Partial Cumulative Model

The constrained partial cumulative model is a nonlinear cumulative probability model that relaxes the parallel regression assumption for two of three subsets of independent variables. In the first group, the coefficients vary freely across the cutpoint equations, as they do in the unconstrained partial model. In the second group, the coefficients are constrained to vary across equations by a set of common factors. In other words, the effects of each variable in this group increases or decreases in strength at the same rate across equations. For example, every coefficient in this group may be 35% weaker in the second cutpoint equation than in the first cutpoint equation. The general form of the equation for the nonlinear cumulative probability model is

$$\Pr(y \leq m \mid \mathbf{x}, \boldsymbol{\omega}, \boldsymbol{\gamma}) = F(\tau_m - \mathbf{x}\boldsymbol{\beta} - \boldsymbol{\omega}\boldsymbol{\eta}_m - \varphi_m \boldsymbol{\gamma}\boldsymbol{\lambda}) \qquad (1 \leq m < M) \tag{3.2}$$

where τ_m are cutpoints, φ_m are common factors $\mathbf{x}$, $\boldsymbol{\omega}$, and $\boldsymbol{\gamma}$ are vectors of independent variables, $\boldsymbol{\beta}$ and $\boldsymbol{\lambda}$ are vectors of coefficients that do not vary across cutpoint equations, and $\boldsymbol{\eta}_m$ is a vector of coefficients that freely vary across cutpoint equations. For the third subset of variables, $\boldsymbol{\gamma}$, the effects vary across equations even though the coefficients are fixed because the overall effect is the product of the common factor and the coefficient: $\varphi_m \lambda$. In order to identify the model, we assume that $\varphi_1 = 1$ and $\varphi_M = 0$ (Fullerton and Xu 2012). We must also assume that $\varphi_1 > \varphi_2 \ldots > \varphi_M$ to ensure the ordinal nature of the relationships. This final constraint is not typically imposed in practice due to estimation difficulties. However, it is possible to impose this constraint within a Bayesian framework (see Chapter 6).

The constrained partial model essentially takes a stereotype logit approach (Anderson 1984) to the third subset of variables. As the number of variables in the third subset and the number of outcome categories increase, the relative parsimony of the constrained partial model compared to the unconstrained model increases as well. For example, if there are two variables in the third subset and two cutpoint equations, the constrained partial model only requires one parameter less than the unconstrained partial model (two slopes and one common factor compared to four slopes). If we increase the number of variables and cutpoint equations from 2 to 4, then the constrained partial model would require nine fewer parameters than the unconstrained partial model (4 slopes and 3 common factors compared to 16 slopes).

Scholars have proposed two conceptual justifications for the common factor constraint. First, the coefficients may vary together across equations for a group of variables because collectively they measure a single, latent construct (Hauser and Andrew 2006). For instance, one may include several indicators of socioeconomic status and therefore expect their effects to vary together across levels of the dependent variable. Second, the coefficients may vary together across equations for a group of binary variables that represent a single, nominal variable (Fullerton and Xu 2012). To illustrate, one might expect the coefficients for binary indicators of religious affiliation (Protestant, Catholic, and other religion) to vary by the same set of factors across cutpoint equations because religious affiliation is expected to get progressively weaker or stronger, overall. Additionally, if one uses the quadratic form to test for nonlinear effects (e.g., age and age-squared), then it is reasonable to expect a common variation pattern across equations.

3.2.3 Example: Partial Cumulative Models of Self-Rated Health

We will illustrate the unconstrained and constrained partial cumulative models using the general self-rated health example from the 2010 GSS. Self-rated health ranges from poor (1) to excellent (4). The independent variables are age (in years), age-squared, sex (female = 1), race/ethnicity (White [ref.], Black, Latino, and other race/ethnicity), education (in years), employment status (employed = 1), log income, and marital status (married = 1). For more details regarding the variables, see Chapter 1.

For the partial models of self-rated health, we relax the parallel regression assumption for age, age-squared, employed, and married and retain the assumption for the remaining variables. We base our decisions regarding the appropriateness of the parallel regression assumption in the examples in this chapter on the amount of coefficient variation across equations. It is appropriate to impose the parallel restriction if the coefficients are relatively constant across equations. We provide an in-depth discussion of tests of the parallel regression assumption in Chapter 5. We present coefficients, odds ratios, and hazard ratios in Table 3.1. The odds and hazard ratios are x-standardized for the continuous variables (age, age-squared, education, and log income). Although negative predicted probabilities are possible in partial cumulative models, all of the observations in our sample have probabilities that fall within the 0–1 range. Therefore, the internal inconsistency does not appear to be a problem in these models.

TABLE 3.1

Results from Unconstrained Partial Cumulative Ordered Regression Models
of Self-Rated Health

	(General Self-Rated Health: 1 = Poor, 2 = Fair, 3 = Good, 4 = Excellent)							
	Logit			Probit		Complementary Log Log		
	Coef.	SE	OR	Coef.	SE	Coef.	SE	HR
Variables								
Age								
1 vs. 2–4	−0.253***	0.061	66.185	−0.118***	0.027	−0.239***	0.058	52.408
1–2 vs. 3–4	−0.086***	0.025	4.149	−0.049***	0.015	−0.068***	0.020	3.060
1–3 vs. 4	−0.065*	0.026	2.946	−0.038*	0.015	−0.038**	0.014	1.870
Age-squared[a]								
1 vs. 2–4	0.215***	0.056	0.027	0.101***	0.025	0.203***	0.053	0.033
1–2 vs. 3–4	0.070**	0.024	0.311	0.040**	0.014	0.055**	0.019	0.398
1–3 vs. 4	0.047#	0.027	0.452	0.028#	0.015	0.029*	0.013	0.620
Female	0.013	0.115	0.987	−0.007	0.067	−0.012	0.073	1.012
Race/ethnicity[b]								
Black	0.093	0.175	0.911	0.030	0.101	0.094	0.112	0.910
Latino	−0.356#	0.191	1.428	−0.218*	0.111	−0.208#	0.119	1.231
Other race/ethnicity	−0.504#	0.290	1.655	−0.284#	0.165	−0.258	0.179	1.295
Education	0.105***	0.022	0.725	0.059***	0.012	0.064***	0.013	0.823
Employed								
1 vs. 2–4	1.625***	0.356	0.197	0.701***	0.159	1.662***	0.344	0.190
1–2 vs. 3–4	0.786***	0.153	0.456	0.457***	0.091	0.704***	0.123	0.495
1–3 vs. 4	0.274#	0.162	0.760	0.158#	0.094	0.137	0.085	0.872
Log income	0.302***	0.062	0.699	0.164***	0.035	0.168***	0.036	0.819
Married								
1 vs. 2–4	0.707*	0.314	0.493	0.357*	0.151	0.784**	0.294	0.457
1–2 vs. 3–4	0.244	0.155	0.783	0.162#	0.090	0.283*	0.124	0.753
1–3 vs. 4	0.045	0.156	0.956	0.035	0.091	0.011	0.084	0.989
Cutpoints								
τ_1	−4.610	1.688		−2.042	0.764	−5.981	1.589	
τ_2	1.565	0.739		0.774	0.430	0.020	0.546	
τ_3	3.863	0.745		2.128	0.432	1.869	0.426	
Model Fit								
R^2_M		0.083			0.080		0.080	
R^2_O		0.108			0.100		0.096	
R^2_N		0.057			0.050		0.051	

Note: The results from significance tests are not reported for the cutpoints. OR = odds ratio, HR = hazard ratio.
  X-standardized odds and hazard ratios are reported for continuous variables (age, age-squared, education,
  and log income), and un-standardized odds and hazard ratios are reported for binary variables
  (female, race/ethnicity, employed, and married). $N = 1129$.
[a] Coefficients and standard errors are multiplied by 100 for ease of presentation.
[b] Reference = White.
#$p < 0.10$, *$p < 0.05$, **$p < 0.01$, ***$p < 0.001$.

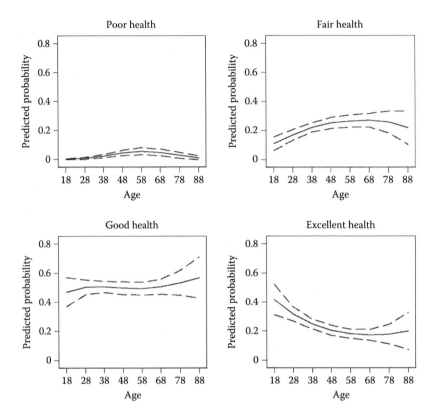

FIGURE 3.1
The effect of age on general self-rated health from an unconstrained partial cumulative logit model.

There are several results that are consistent across the three models. The coefficients for age and age-squared are significant in each model. We will focus on these nonlinear patterns in more detail in Figure 3.1. Education and log income have significant, positive effects on health and the coefficients are constrained to remain constant across levels of health. Using the logit link, we see that a standard deviation increase in education is associated with a decrease in the odds of poorer health by a factor of 0.73 or 27%.* Similarly, a standard deviation increase in the log of family income is associated with a decrease in the odds of poorer health by a factor of 0.70 or 30%.† Sex differences in health are nonsignificant, but the differences by race and ethnicity are at least marginally significant ($p < 0.10$).

We relax the parallel regression assumption for the remaining variables in the models. The coefficients are consistently larger at lower levels of health. For example, employment is associated with a decrease in the hazard rate of poor health by a factor of 0.19 or 81%, but the hazard rates for poor to fair health and poor to good health are 0.50 (or 50%) and 0.87 (or 13%), respectively. In other words, having a job reduces the risk of being in poor health, but it does not guarantee that one will be in excellent health. A similar pattern emerges for marital status. Being married is associated with a decrease in the odds of poor health by

---

* Interpretations of odds/hazard ratios assume that the remaining variables in the model are held constant.
† Recall that due to the parameterization of the models as $\tau - x\beta$, odds and hazard ratios are calculated as $\exp(-\beta)$ for unstandardized factor changes and $\exp(-\beta*S_x)$ for x-standardized factor changes, where S_x is the standard deviation of x.

a factor of 0.49 or 51%, but the odds ratios for poor to fair health and poor to good health are 0.78 (or 22%) and 0.96 (or 4%), respectively. Once again, a lower odds of poor health does not correspond with a higher odds of excellent health. Relaxing the parallel regression assumption allows us to observe these "asymmetrical" patterns (Fullerton and Dixon 2009, 2010).

According to the model fit statistics, this group of variables reduces the model deviance by 8%, based on the R_M^2. The model fit is best according to the ordinal dispersion measure. The models explain 10%–11% of the ordinal variation, compared to only 5%–6% of the nominal variation. These results provide further support for the notion that partial models allow one to estimate ordinal relationships even though the strict stochastic ordering is only guaranteed for the variables that retain the parallel regression assumption.

In Table 3.2, we present coefficients, odds ratios, and hazard ratios from constrained partial cumulative models of self-rated health. The key difference between the unconstrained and constrained partial models is that the coefficients for age and age-squared vary across the cutpoint equations by a set of common factors in the constrained model. There are three cutpoint equations, which means there are two common factors (φ_2 and φ_3). In order to identify the model, we assume that $\varphi_1 = 1$ and $\varphi_4 = 0$. For the constrained partial models, we test the statistical significance of the common factors using a Wald test based on the following null hypothesis: $\varphi_m = 1$. A value of 1 for φ_2 would indicate that the effects are the same in Cutpoint Equations 1 and 2.

Age and age-squared have significant coefficients. The probit results for the common factors indicate that these coefficients are 54% weaker in the second cutpoint equation and 57% weaker in the third cutpoint equation compared to the first equation. The changes are even greater using the complementary log log model (69% and 80%). The results are very similar for the constrained and unconstrained partial models of self-rated health, but the constrained model is slightly more parsimonious due to the use of common factors for age and age-squared (four vs. six parameters for age and age-squared).

We examine the substantive significance of the findings from the partial cumulative logit models using average marginal effects (AMEs), which we present in Table 3.3. For continuous variables, we also include average discrete changes (ADCs) corresponding to standard deviation increases in parentheses. The AMEs are very similar for the unconstrained and constrained partial models. The main difference is that the effects of age and age-squared are more consistently significant across categories of self-rated health in the constrained partial model, which is due to the use of common factors to allow for variation across levels of health. Several variables have stronger effects on poor health than on excellent health, including employment and marital status. On average, being married significantly reduces the probability of poor health, but it does not have a significant effect on fair, good, or excellent health. Employment status has more consistent effects across equations, but its effect on the probability of excellent health is weaker and only marginally significant ($p < 0.10$). However, variables that retain the parallel regression assumption, such as education and income, actually have stronger effects on excellent health than any other category. The parallel regression assumption has clear implications for category-specific tests of statistical significance, which makes decisions regarding whether to retain or relax the assumption very important. For an in-depth discussion of those decisions regarding the parallel assumption, see Chapter 5.

We examine the nonlinear effects of age on self-rated health by plotting the predicted probabilities across the range of age in Figure 3.1.* All else equal, young respondents

---

* The predicted probabilities and 95% confidence intervals in this chapter are based on the unconstrained partial models using the logit link with the remaining variables held constant at their means.

TABLE 3.2

Results from Constrained Partial Cumulative Ordered Regression Models of Self-Rated Health

	(General Self-Rated Health: 1 = Poor, 2 = Fair, 3 = Good, 4 = Excellent)							
	Logit			Probit		Complementary Log Log		
	Coef.	SE	OR	Coef.	SE	Coef.	SE	HR
Variables								
Female	0.013	0.115	0.987	−0.008	0.067	−0.014	0.073	1.015
Race/ethnicity[b]								
Black	0.108	0.174	0.897	0.037	0.101	0.105	0.112	0.900
Latino	−0.347[#]	0.191	1.415	−0.214[#]	0.111	−0.200[#]	0.119	1.222
Other race/ethnicity	−0.497[#]	0.290	1.644	−0.280[#]	0.165	−0.249	0.179	1.282
Education	0.105***	0.022	0.725	0.059***	0.012	0.064***	0.013	0.822
Employed								
1 vs. 2–4	1.580***	0.354	0.206	0.674***	0.157	1.620***	0.342	0.198
1–2 vs. 3–4	0.800***	0.148	0.449	0.465***	0.088	0.714***	0.119	0.490
1–3 vs. 4	0.330*	0.156	0.719	0.189*	0.091	0.163*	0.082	0.850
Log income	0.305***	0.062	0.696	0.165***	0.035	0.168***	0.036	0.819
Married								
1 vs. 2–4	0.701*	0.314	0.496	0.356*	0.151	0.780**	0.295	0.458
1–2 vs. 3–4	0.247	0.155	0.781	0.164[#]	0.090	0.287*	0.124	0.751
1–3 vs. 4	0.058	0.155	0.943	0.043	0.090	0.018	0.084	0.982
Variables with Partial Proportionality Constraints (PPC)								
Age	−0.241***	0.061	53.793	−0.112***	0.028	−0.227***	0.059	43.023
Age-squared[a]	0.199***	0.055	0.035	0.093***	0.025	0.188***	0.053	0.042
PPC								
φ_2	0.375***	0.104		0.462***	0.125	0.310***	0.088	
φ_3	0.348***	0.108		0.429***	0.131	0.200***	0.064	
Cutpoints								
τ_1	−4.408	1.703		−1.945	0.774	−5.805	1.606	
τ_2	1.526	0.720		0.745	0.417	−0.012	0.528	
τ_3	3.606	0.700		1.984	0.408	1.765	0.412	
Model Fit								
R^2_M		0.083			0.080		0.079	
R^2_O		0.108			0.100		0.095	
R^2_N		0.056			0.050		0.051	

Note: OR = odds ratio, HR = hazard ratio. X-standardized odds and hazard ratios are reported for continuous variables (age, age-squared, education, and log income), and un-standardized odds and hazard ratios are reported for binary variables (female, race/ethnicity, employed, and married).

[a] Coefficients and standard errors are multiplied by 100 for ease of presentation.

[b] Reference = White.

[#]$p < 0.10$, *$p < 0.05$, **$p < 0.01$, ***$p < 0.001$.

tend to rate their health better than older respondents. From age 18 to 58, an increase in age is associated with poorer self-rated health. The predicted probability of poor health increases from 0.002 to 0.057, and the probability of fair health increases from 0.111 to 0.267. The predicted probability of excellent health dramatically declines from 0.417 to 0.182. However, after age 68, increases in age are associated with modest increases in health.

TABLE 3.3

AMEs from Unconstrained and Constrained Partial Cumulative Logit Models of Self-Rated Health

	Poor	Fair	Good	Excellent
Unconstrained Partial Model				
Age	0.012***	0.004	−0.004	−0.011*
	(0.524)	(−0.234)	(−0.146)	(−0.144)
Age-squared[a]	−0.010***	−0.002	0.004	0.008#
	(−0.053)	(−0.107)	(0.001)	(0.159)
Female	−0.001	−0.002	0.000	0.002
Race/ethnicity[b]				
Black	−0.004	−0.012	0.000	0.016
Latino	0.018#	0.048#	−0.007	−0.059*
Other race/ethnicity	0.027	0.068#	−0.016	−0.079*
Education	−0.005***	−0.014***	0.000	0.018***
	(−0.013)	(−0.040)	(−0.007)	(0.060)
Employed	−0.069***	−0.077**	0.099**	0.047#
Log income	−0.014***	−0.039***	0.000	0.053***
	(−0.014)	(−0.044)	(−0.008)	(0.067)
Married	−0.030*	−0.013	0.035	0.008
Constrained Partial Model				
Age	0.011***	0.005***	−0.001	−0.015***
	(0.487)	(−0.182)	(−0.136)	(−0.169)
Age-squared[a]	−0.009***	−0.004***	0.001	0.012***
	(−0.052)	(−0.116)	(−0.073)	(0.241)
Female	−0.001	−0.002	0.000	0.002
Race/ethnicity[b]				
Black	−0.005	−0.014	−0.001	0.019
Latino	0.017#	0.047#	−0.007	−0.057#
Other race/ethnicity	0.027	0.067#	−0.015	−0.078#
Education	−0.005***	−0.014***	0.000	0.018***
	(−0.013)	(−0.040)	(−0.007)	(0.060)
Employed	−0.067***	−0.082**	0.092**	0.057*
Log income	−0.014***	−0.040***	0.000	0.053***
	(−0.015)	(−0.045)	(−0.009)	(0.068)
Married	−0.030*	−0.013	0.033	0.010

Note: The numbers in parentheses are average standard deviation discrete changes.

[a] Coefficient multiplied by 100 for ease of presentation.

[b] Reference = White.

#$p < 0.10$, *$p < 0.05$, **$p < 0.01$, ***$p < 0.001$.

The predicted probability of excellent health increases to 0.20, and the predicted probability of poor or fair health decreases to 0.01 and 0.22, respectively. The probability of good health is relatively constant throughout the age distribution, but the probability does increase slightly from age 18 to 28 (0.47 to 0.50) and then again from age 68 to 88 (0.50 to 0.57). However, the confidence bands are also wider at the ends of the age distribution due to the smaller sample sizes.

3.3 Partial Continuation Ratio Models

Partial continuation ratio models are an extension of the parallel continuation ratio model that relaxes the parallel regression assumption for a subset of variables. Fortunately, partial continuation ratio models do not have the same internal inconsistency found in partial cumulative models. The sequential nature of the data-generating process and progressively smaller sample sizes in later cutpoint equations ensure that the predicted probabilities will be nonnegative. However, the conditional nature of this process poses other problems.

Stage models essentially select on the dependent variable. Only respondents that "survive" earlier transitions are included in subsequent stages or cutpoint equations. As a result, the sample size is smaller in later stages, and the residual variance may decrease as well if there is more homogeneity in later stages. Increasing sample homogeneity in later stages is a sample selection effect that may potentially lead to biased coefficients in those stages. Additionally, the coefficients in nonlinear probability models are standardized by the residual variation and only identified up to a scale based on assumptions about the variance (Mood 2010). Therefore, variation in coefficients across cutpoint equations may reflect differences in residual variation or differences in the "true effects" of independent variables. It is impossible to distinguish between changes in the effects of variables and changes in the residual variances across stages (Mare 2006, pp. 30–31).

Fortunately, AMEs are not sensitive to changes in the scale of the latent outcome, but they are potentially biased by selection on the dependent variable (for more details, see Cameron and Heckman 1998). As a result, we will focus more on variation in AMEs than on odds/hazard ratios or raw coefficients across cutpoint equations, and we will keep in mind that variation across equations could be due to either changes in "true effects" or to heteroscedastic errors. Williams (2010) notes that heterogeneous choice models, which account for heteroscedastic errors in nonlinear probability models, may be a reasonable alternative when the parallel regression assumption is violated. For a discussion of these models, see Chapter 6.

3.3.1 Unconstrained Partial Continuation Ratio Models

The unconstrained partial continuation ratio model is a nonlinear conditional probability model that relaxes the parallel regression assumption for one subset of variables. The probability is conditional on making the transition to that particular stage. The general form of the equation for the nonlinear conditional probability model is

$$\Pr(y = m \mid y \geq m, \mathbf{x}, \boldsymbol{\omega}) = F(\tau_m - \mathbf{x}\boldsymbol{\beta} - \boldsymbol{\omega}\boldsymbol{\eta}_m) \qquad (1 \leq m < M) \qquad (3.3)$$

where F typically represents the CDF for the logit, probit, or complementary log log model; τ_m are cutpoints; $\mathbf{x}$ and $\boldsymbol{\omega}$ are vectors of independent variables; $\boldsymbol{\beta}$ is a vector of coefficients that do not vary across cutpoint equations; and $\boldsymbol{\eta}_m$ is a vector of coefficients that freely vary across cutpoint equations.

The original, ordinal outcome is recoded into a series of $M - 1$ binary outcomes corresponding to each cutpoint equation. If $M = 4$, then the unconstrained partial continuation ratio model estimates the following three binary cutpoint equations simultaneously:

$$\text{Cutpoint equation \#1: } \Pr(y = 1 \mid y \geq 1, \mathbf{x}, \boldsymbol{\omega}) = F(\tau_1 - \mathbf{x}\boldsymbol{\beta} - \boldsymbol{\omega}\boldsymbol{\eta}_1)$$
$$\text{Cutpoint equation \#2: } \Pr(y = 2 \mid y \geq 2, \mathbf{x}, \boldsymbol{\omega}) = F(\tau_2 - \mathbf{x}\boldsymbol{\beta} - \boldsymbol{\omega}\boldsymbol{\eta}_2)$$
$$\text{Cutpoint equation \#3: } \Pr(y = 3 \mid y \geq 3, \mathbf{x}, \boldsymbol{\omega}) = F(\tau_3 - \mathbf{x}\boldsymbol{\beta} - \boldsymbol{\omega}\boldsymbol{\eta}_3)$$

The values for the cutpoints (τ_m) and the coefficients for the second set of independent variables (η_m) vary across equations, whereas the values for the independent variables (**x** and **ω**) and the coefficients for the first set of variables (**β**) remain constant across equations. This partial model is unconstrained in the sense that the values for η_m freely vary across cutpoint equations. In the constrained partial model, these coefficients vary by one or more common factors across equations.

3.3.2 Constrained Partial Continuation Ratio Models

The constrained partial continuation ratio model is a nonlinear conditional probability model that relaxes the parallel regression assumption for two subsets of independent variables in the model. It allows variables to either have coefficients that are fixed, freely varying, or varying by common factors across cutpoint equations. The general form of the equation for the nonlinear conditional probability model is

$$\Pr(y = m \mid y \geq m, \mathbf{x}, \boldsymbol{\omega}, \boldsymbol{\gamma}) = F(\tau_m - \mathbf{x}\boldsymbol{\beta} - \boldsymbol{\omega}\eta_m - \varphi_m\boldsymbol{\gamma}\lambda) \qquad (1 \leq m < M) \tag{3.4}$$

where τ_m are cutpoints; φ_m are common factors; **x**, **ω**, and **γ** are vectors of independent variables; **β** and λ are vectors of coefficients that do not vary across cutpoint equations; and η_m is a vector of coefficients that freely vary across cutpoint equations. Hauser and Andrew (2006) introduced this model in the context of educational transitions and referred to it as the logistic response with partial proportionality constraints model.

3.3.3 Example: Partial Continuation Ratio Models of Educational Attainment

Partial continuation ratio models are well suited to analyze educational attainment because the attainment process is a sequence of irreversible stages and the models are flexible enough to allow some variables to have effects that vary across transitions. The data for this example come from the 2010 GSS. The dependent variable, educational attainment, is a five-category ordinal measure ranging from less than high school (1) to postgraduate (5). The independent variables are age (in years), sex (female = 1), race/ethnicity (White [ref.], Black, Latino, and other race/ethnicity), and parents' education (mother's and father's education in years). For more details regarding the variables, see Chapter 1.

In Table 3.4, we present coefficients and odds/hazard ratios from unconstrained partial continuation ratio models. There are five outcome categories, which means there are four cutpoint equations or educational stages. We retain the parallel regression assumption for sex and race/ethnicity and relax it for age and parents' education. Sex differences in educational attainment are not statistically significant, but there is at least one significant racial difference. For non-Hispanic respondents, identifying with a race other than White or Black (compared to White) is associated with a decrease in the odds of a lower level of educational attainment by a factor of 0.45 or 55%.

The coefficients for age and the two measures of parents' education are allowed to freely vary across the educational transitions. Age does not have a significant effect in the first stage (less than high school vs. high school or more), but it does reduce the odds of lower attainment levels in the subsequent stages. For example, a standard deviation increase in age is associated with a decrease in the hazard rate for a college degree (vs. postgraduate) by a factor of 0.82 or 18%. The coefficients for the two measures of parents' education are at least marginally significant ($p < 0.10$) in each stage, but the coefficients are larger and more consistently significant in the earlier stages. For example, a standard

TABLE 3.4

Results from Unconstrained Partial Continuation Ratio Ordered Regression Models of Educational Attainment

	(Educational Attainment: 1 = Less than High School, 2 = High School, 3 = Associate, 4 = College, 5 = Postgraduate)							
	Logit			Probit			Complementary Log Log		
	Coef.	SE	OR	Coef.	SE	Coef.	SE	HR
Variables								
Age								
1 vs. 2–5	−0.004	0.007	1.066	−0.001	0.004	−0.003	0.006	1.049
2 vs. 3–5	0.011**	0.004	0.835	0.007*	0.003	0.007*	0.003	0.901
3 vs. 4–5	0.027**	0.008	0.653	0.015***	0.005	0.023**	0.007	0.694
4 vs. 5	0.016*	0.007	0.772	0.010*	0.004	0.012**	0.004	0.823
Female	0.101	0.088	0.904	0.056	0.051	0.069	0.063	0.933
Race/ethnicity[a]								
Black	−0.221	0.154	1.248	−0.135	0.089	−0.099	0.108	1.104
Latino	−0.275	0.178	1.317	−0.159	0.103	−0.118	0.123	1.125
Other race/ethnicity	0.798***	0.232	0.450	0.457***	0.134	0.606***	0.183	0.546
Parents' education								
Mother's education								
1 vs. 2–5	0.229***	0.037	0.422	0.127***	0.020	0.192***	0.031	0.486
2 vs. 3–5	0.084**	0.026	0.729	0.047**	0.016	0.039*	0.017	0.864
3 vs. 4–5	0.100*	0.046	0.686	0.057*	0.025	0.082*	0.040	0.734
4 vs. 5	0.066#	0.040	0.780	0.041#	0.024	0.043#	0.024	0.850
Father's education								
1 vs. 2–5	0.115***	0.034	0.614	0.062***	0.018	0.097***	0.029	0.662
2 vs. 3–5	0.138***	0.022	0.557	0.083***	0.013	0.086***	0.015	0.695
3 vs. 4–5	0.078#	0.040	0.720	0.048*	0.022	0.059#	0.035	0.779
4 vs. 5	−0.061#	0.033	1.296	−0.037#	0.020	−0.036#	0.020	1.167
Cutpoints								
τ_1	0.914	0.529		0.533	0.286	0.359	0.404	
τ_2	3.256	0.432		1.896	0.250	1.450	0.262	
τ_3	1.948	0.724		1.127	0.411	1.194	0.547	
τ_4	1.505	0.654		0.947	0.400	0.759	0.411	
Model Fit								
R^2_M		0.110			0.110		0.103	
R^2_O		0.145			0.141		0.121	
R^2_N		0.083			0.078		0.070	

Note: The results from significance tests are not reported for the cutpoints. OR = odds ratio, HR = hazard ratio. X-standardized odds and hazard ratios are reported for continuous variables (age and parents' education), and un-standardized odds and hazard ratios are reported for binary variables (female and race/ethnicity). $N = 1286$.

[a] Reference = White.

#$p < 0.10$, *$p < 0.05$, **$p < 0.01$, ***$p < 0.001$.

deviation increase in mother's education is associated with a larger decrease in the hazard rate for less than a high school degree (vs. high school or more) than a college degree (vs. postgraduate). Additionally, the coefficient for father's education actually changes sign in the final stage.

The model fit measures indicate that this group of variables reduces the model deviance by 10%–11%. In addition, these variables explain 12%–15% of the ordinal variation and 7%–8% of the nominal variation in educational attainment. Finally, the logit and probit models tend to provide a slightly better fit to the data than the complementary log log model.

In Table 3.5, we present the results from constrained partial continuation ratio models of educational attainment. We constrain the coefficients for mother's and father's education to vary by common factors across educational transitions. The results for age, sex, and race/ethnicity are very similar in the unconstrained and constrained partial models. The constrained partial model is more parsimonious than the unconstrained partial model because it only requires five parameters for parents' education, whereas the unconstrained model required eight. In the first stage, a standard deviation increase in father's education is associated with a decrease in the odds of attaining less than a high school degree by a factor of 0.50 or 50%. The common factors (φ) indicate that this relationship gets weaker in the second and third educational stages and then changes direction in the final stage. We assume that $\varphi_1 = 1$ and $\varphi_5 = 0$ in order to identify the models. The model does not actually constrain $\varphi_2 - \varphi_4$ to fall within the 0–1 range. We can calculate the odds ratios for Stages 2 through 4 by taking the exponent of product of the slope and common factor: OR = $\exp(-\varphi_m \lambda)$. The x-standardized odds ratios for father's education in Stages 2 through 4 are

$$\text{Stage 2: } \exp(-1 * 0.671 * 4.235 * 0.162) = 0.631$$

$$\text{Stage 3: } \exp(-1 * 0.518 * 4.235 * 0.162) = 0.701$$

$$\text{Stage 4: } \exp(-1 * -0.028 * 4.235 * 0.162) = 1.019$$

Although φ_4 is negative, the odds ratio in Stage 4 indicates that there is little to no effect in this stage. The odds ratios are very similar for mother's education across the educational stages.

We present the AMEs in Table 3.6 based on the partial continuation ratio logit models to examine the substantive significance of these findings. The ADCs for continuous variables are also included in parentheses. This is particularly important given the fact that changes in raw coefficients or odds/hazard ratios may be due to unobserved heterogeneity in sequential models (Cameron and Heckman 1998; Buis 2011). The AMEs are very similar in the unconstrained and constrained partial continuation ratio logit models. However, the common factor constraints in the constrained partial model produce more consistently significant effects for parents' education and substantially different AMEs in the last stage. Racial differences are most pronounced in the high school and college stages, and age differences appear to be more important in later educational stages.

The AMEs are relatively constant across stages for mother's education in the unconstrained partial model. The average effect on the predicted probability is only marginally significant in the college stage, but this is likely due to the smaller sample size in this cutpoint equation. Only 38% of the respondents made the earlier transitions to the college versus postgraduate stage. However, the AMEs for father's education change

TABLE 3.5

Results from Constrained Partial Continuation Ratio Ordered Regression Models of Educational Attainment

	Logit			Probit		Complementary Log Log		
(Educational Attainment: 1 = Less than High School, 2 = High School, 3 = Associate, 4 = College, 5 = Postgraduate)								
	Coef.	SE	OR	Coef.	SE	Coef.	SE	HR
Variables								
Age								
1 vs. 2–5	−0.004	0.007	1.062	−0.001	0.004	−0.003	0.006	1.049
2 vs. 3–5	0.011**	0.004	0.833	0.007*	0.003	0.006*	0.003	0.902
3 vs. 4–5	0.027**	0.008	0.654	0.015***	0.005	0.023**	0.007	0.693
4 vs. 5	0.015*	0.007	0.785	0.010*	0.004	0.011**	0.004	0.835
Female	0.107	0.088	0.898	0.060	0.051	0.071	0.063	0.931
Race/ethnicity[a]								
Black	−0.219	0.153	1.245	−0.133	0.089	−0.096	0.108	1.101
Latino	−0.298#	0.178	1.347	−0.170#	0.103	−0.117	0.122	1.124
Other race/ethnicity	0.803***	0.232	0.448	0.457***	0.134	0.615***	0.183	0.541
Variables with Partial Proportionality Constraints (PPC)								
Parents' education								
Mother's education	0.178***	0.029	0.512	0.098***	0.016	0.149***	0.025	0.571
Father's education	0.162***	0.024	0.503	0.089***	0.013	0.140***	0.022	0.553
PPC								
φ_2	0.671***	0.090		0.718**	0.093	0.445***	0.054	
φ_3	0.518***	0.118		0.563***	0.123	0.485***	0.100	
φ_4	−0.028***	0.095		−0.027***	0.106	−0.010***	0.069	
Cutpoints								
τ_1	0.881	0.531		0.516	0.288	0.344	0.406	
τ_2	3.348	0.428		1.949	0.248	1.494	0.259	
τ_3	1.930	0.723		1.123	0.410	1.186	0.547	
τ_4	1.277	0.643		0.806	0.395	0.599	0.409	
Model Fit								
R^2_M		0.107			0.108		0.100	
R^2_O		0.143			0.139		0.119	
R^2_N		0.079			0.074		0.066	

Note: SE = standard error, OR = odds ratio, HR = hazard ratio. X-standardized odds and hazard ratios are reported for continuous variables (age and parents' education), and un-standardized odds and hazard ratios are reported for binary variables (female and race/ethnicity). $N = 1286$.

[a] Reference = White.

#$p < 0.10$, *$p < 0.05$, **$p < 0.01$, ***$p < 0.001$.

considerably across the transitions. The average effect on the predicted probability is similar in the first and third stages, but there is a stronger, negative effect in the high school stage and then it changes direction in the final stage. Father's education is associated with an increase in the predicted probability of a college degree (vs. a postgraduate degree), whereas mother's education is associated with a decrease in this probability.

TABLE 3.6

Conditional AMEs from Unconstrained and Constrained Partial Continuation Ratio Logit Models of Educational Attainment

	Less than High School	High School	Associate	College
Unconstrained Partial Model				
Age	0.000	−0.002**	−0.004***	−0.004*
	(0.004)	(−0.039)	(−0.056)	(−0.060)
Female	−0.007	−0.022	−0.015	−0.023
Race/ethnicity[a]				
Black	0.015	0.048	0.034	0.048
Latino	0.020	0.060	0.043	0.060
Other race/ethnicity	−0.043***	−0.169***	−0.096***	−0.189***
Parents' education				
Mother's education	−0.015***	−0.018**	−0.015*	−0.015#
	(−0.045)	(−0.068)	(−0.050)	(−0.057)
Father's education	−0.008***	−0.030***	−0.011#	0.014#
	(−0.028)	(−0.125)	(−0.044)	(0.056)
Constrained Partial Model				
Age	0.000	−0.002**	−0.004***	−0.003*
	(0.004)	(−0.040)	(−0.056)	(−0.056)
Female	−0.007	−0.023	−0.016	−0.024
Race/ethnicity[a]				
Black	0.015	0.048	0.034	0.048
Latino	0.022	0.065#	0.047	0.065#
Other race/ethnicity	−0.044***	−0.170***	−0.097***	−0.192***
Parents' education				
Mother's education	−0.012***	−0.026***	−0.014***	0.001***
	(−0.037)	(−0.097)	(−0.047)	(0.004)
Father's education	−0.011***	−0.024***	−0.012***	0.001***
	(−0.038)	(−0.099)	(−0.048)	(0.004)

Note: The numbers in parentheses are average standard deviation discrete changes.

[a] Reference = White.

#$p < 0.10$, *$p < 0.05$, **$p < 0.01$, ***$p < 0.001$.

In Figure 3.2, we present predicted probabilities across the range of mother's education to examine its association with educational attainment in more detail. The confidence intervals are widest between zero and seven years for mother's education because only 11% of respondents have mothers with less than 8 years of schooling. An increase from 0 to 20 years of mother's education is associated with a substantial increase in the conditional predicted probability in each stage. Increases in mother's education are associated with decreases in the probabilities of lower educational levels in each stage. The strongest effect is in the first stage. For respondents that have mothers with no formal schooling, the predicted probability of attaining less than a high school degree is 0.44. However, for respondents that have mothers with 20 years of schooling, the predicted probability of attaining less than a high school degree is 0.01.

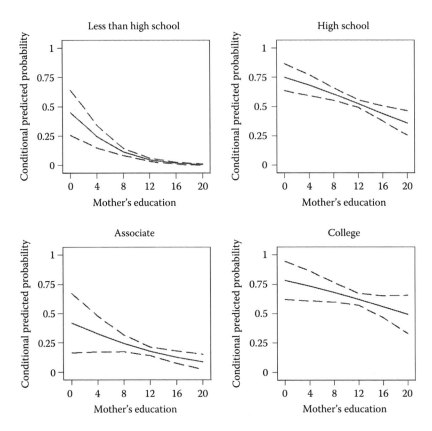

FIGURE 3.2
The conditional effect of mother's education on educational attainment from an unconstrained partial continuation ratio logit model.

3.4 Partial Adjacent Category Models

Partial adjacent category models are an extension of the parallel adjacent category model that relaxes the parallel regression assumption for a subset of variables. Additionally, the partial adjacent category logit model is a constrained version of the more familiar multinomial logit model. Partial adjacent category models do not have the same internal inconsistency found in partial cumulative models. However, like the stage models, the adjacent models essentially select on the dependent variable. The cutpoint equations only involve respondents in two adjacent categories. As a result, adjacent models must rely on an additional assumption, the independence of irrelevant alternatives (IIA), which requires that the ratio of the probabilities for two categories is not affected by the addition of new alternatives (Long 1997, pp. 182–183). In order to use adjacent models, researchers should be confident that the outcome categories are distinct and weighed independently by respondents (McFadden 1973). For more details regarding the IIA assumption, see the discussion in Section 2.3.1.

Partial adjacent category logit models are also similar to Anderson's (1984) single-dimension stereotype logit model, which assumes that the coefficients for every independent variable in the model vary across equations by a set of common factors. The stereotype logit model is a vector generalized linear model and a reduced rank version of the well-known multinomial logit model (Yee and Hastie 2003; Yee 2015). The equation for the log odds in the single-dimension stereotype logit model is

$$\ln\left[\frac{\Pr(y=m\mid\mathbf{x})}{\Pr(y=M\mid\mathbf{x})}\right]=\tau_m-\varphi_m\mathbf{x}\boldsymbol{\beta} \tag{3.5}$$

where τ_m are cutpoints, φ_m are common factors, $\mathbf{x}$ is a vector of independent variables, and $\boldsymbol{\beta}$ is a vector of coefficients. The stereotype model is set up as a baseline logit model—like multinomial logit—but one could easily set it up using adjacent comparisons as well. The stereotype model combines the parsimony of the parallel model with the flexibility of the nonparallel model. The partial adjacent category logit models would simplify to the one-dimensional stereotype logit model if we constrained every independent variable to have coefficients that vary by common factors.

3.4.1 Unconstrained Partial Adjacent Category Models

The unconstrained partial adjacent category model is a nonlinear adjacent probability model that relaxes the parallel regression assumption for one subset of variables. The "adjacent probability" (Goodman 1983) is a conditional probability because the focus is limited to two adjacent categories. The general form of the equation for the nonlinear adjacent probability model is

$$\Pr(y=m\mid y=m\text{ or }y=m+1,\mathbf{x},\boldsymbol{\omega})=F(\tau_m-\mathbf{x}\boldsymbol{\beta}-\boldsymbol{\omega}\boldsymbol{\eta}_m) \qquad (1\le m<M) \tag{3.6}$$

where F typically represents the CDF for the logit or probit model,[*] τ_m are cutpoints, $\mathbf{x}$ and $\boldsymbol{\omega}$ are vectors of independent variables, $\boldsymbol{\beta}$ is a vector of coefficients that do not vary across cutpoint equations, and $\boldsymbol{\eta}_m$ is a vector of coefficients that freely vary across cutpoint equations.

The original, ordinal outcome is recoded into a series of $M-1$ binary outcomes corresponding to each cutpoint equation. If $M=4$, then the unconstrained partial adjacent category model estimates the following three binary cutpoint equations simultaneously:

Cutpoint equation #1: $\Pr(y=1\mid y=1\text{ or }y=2,\mathbf{x},\boldsymbol{\omega})=F(\tau_1-\mathbf{x}\boldsymbol{\beta}-\boldsymbol{\omega}\boldsymbol{\eta}_1)$

Cutpoint equation #2: $\Pr(y=2\mid y=2\text{ or }y=3,\mathbf{x},\boldsymbol{\omega})=F(\tau_2-\mathbf{x}\boldsymbol{\beta}-\boldsymbol{\omega}\boldsymbol{\eta}_2)$

Cutpoint equation #3: $\Pr(y=3\mid y=3\text{ or }y=4,\mathbf{x},\boldsymbol{\omega})=F(\tau_3-\mathbf{x}\boldsymbol{\beta}-\boldsymbol{\omega}\boldsymbol{\eta}_3)$

The values for the cutpoints (τ_m) and the coefficients for the second set of independent variables ($\boldsymbol{\eta}_m$) vary across equations, whereas the values for the independent variables ($\mathbf{x}$ and $\boldsymbol{\omega}$) and the coefficients for the first set of variables ($\boldsymbol{\beta}$) remain constant across equations. There are no constraints placed on the manner in which the coefficients in $\boldsymbol{\eta}_m$ vary across equations, which is what distinguishes the unconstrained and constrained partial models.

---

[*] The probit version of the partial adjacent category model also requires the IIA assumption. We do not consider the alternative-specific multinomial probit model, which allows for correlated errors across equations (see Long and Freese 2014, p. 469).

3.4.2 Constrained Partial Adjacent Category Models

The constrained partial adjacent category model is a nonlinear adjacent probability model that relaxes the parallel regression assumption for two subsets of independent variables in the model. It allows variables to either have coefficients that are fixed, freely varying, or varying by common factors across cutpoint equations. The general form of the equation for the nonlinear adjacent probability model is

$$\Pr(y = m \mid y = m \text{ or } y = m + 1, \mathbf{x}, \boldsymbol{\omega}, \boldsymbol{\gamma}) = F(\tau_m - \mathbf{x}\boldsymbol{\beta} - \boldsymbol{\omega}\boldsymbol{\eta}_m - \varphi_m\boldsymbol{\gamma}\lambda) \qquad (1 \le m < M) \qquad (3.7)$$

where τ_m are cutpoints; φ_m are common factors; $\mathbf{x}$, $\boldsymbol{\omega}$, and $\boldsymbol{\gamma}$ are vectors of independent variables; $\boldsymbol{\beta}$ and λ are vectors of coefficients that do not vary across cutpoint equations; and $\boldsymbol{\eta}_m$ is a vector of coefficients that freely vary across cutpoint equations. Fullerton (2009) introduced this model, and Fullerton and Xu (2015) discussed its connection to other adjacent category models. The more familiar stereotype logit model (see Anderson 1984; Greenland 1994) is a constrained form of this model in which we assume that $\boldsymbol{\beta} = 0$ and $\boldsymbol{\eta}_m = 0$.

3.4.3 Example: Partial Adjacent Category Models of Welfare Spending Attitudes

Adjacent category models are preferable over cumulative and stage models when the outcome categories are of substantive interest and the assumption of a sequence of irreversible stages is not realistic. Adjacent category models are well suited for government spending questions because the categories reflect three distinct spending views, but the assumption of a sequential decision-making process is not applicable. The data for this example come from the 2012 GSS. The dependent variable, welfare spending attitude, is a three-category ordinal measure ranging from too much (1) to too little (3). The independent variables are age, sex, income, education, ideology, party identification, and negative racial attitude. In accordance with previous research (e.g., Gilens 1999), we limit the sample to White respondents given the racialized nature of welfare attitudes in the United States. For more details regarding the variables, see Chapter 1.

In Table 3.7, we present coefficients and odds ratios from unconstrained partial adjacent category models using logit and probit link functions. There are three outcome categories, which means there are two cutpoint equations. We retain the parallel regression assumption for sex, income, and racial attitude, and we relax the assumption for age, education, ideology, and party identification. Sex differences in welfare attitudes are not statistically significant, but income and negative racial attitudes are associated with higher levels of opposition to welfare spending. For example, a standard deviation increase in log income is associated with an increase in the odds of greater opposition to welfare spending by a factor of 1.31 or 31%.

The coefficients for the remaining variables are free to vary across the cutpoint equations, and this variation is substantial. The coefficient for age is positive and significant in the first equation (too much vs. about right) but negative and nonsignificant in the second equation (about right vs. too little). A standard deviation increase in age is associated with a reduction in the odds of opposition to welfare spending ("too much") by a factor of 0.76 or 24%, but this does not coincide with an increase in the odds of support ("too little"). We observe a similar asymmetrical pattern for education. Education does not have a significant effect on the odds of opposition to welfare spending, but it is associated with an increase in the odds of mixed support ("about right") relative to support. All else equal, more educated respondents are more likely to take the mixed view of welfare spending, and education appears to have a larger relative effect on support for welfare spending than opposition to it.

TABLE 3.7

Results from Unconstrained Partial Adjacent Category Ordered Regression Models of Welfare Spending Attitudes

	(Welfare Spending Attitude: 1 = Too Much, 2 = About Right, 3 = Too Little)				
	Logit			Probit	
	Coef.	SE	OR	Coef.	SE
Variables					
Age					
1 vs. 2	0.016*	0.007	0.763	0.012*	0.006
2 vs. 3	−0.004	0.009	1.067	−0.003	0.007
Female	−0.017	0.151	1.017	0.005	0.114
Income	−0.237***	0.070	1.311	−0.185***	0.053
Education					
1 vs. 2	0.075	0.046	0.813	0.059	0.037
2 vs. 3	−0.134*	0.059	1.451	−0.106*	0.043
Ideology					
1 vs. 2	−0.011	0.097	1.017	−0.012	0.079
2 vs. 3	−0.221#	0.125	1.377	−0.166#	0.091
Party identification[a]					
Democrat					
1 vs. 2	1.253***	0.364	0.286	1.040***	0.293
2 vs. 3	0.414	0.565	0.661	0.172	0.381
Independent					
1 vs. 2	1.133***	0.327	0.322	0.929***	0.262
2 vs. 3	0.412	0.539	0.662	0.195	0.356
Racial attitude	−0.239**	0.077	1.273	−0.181**	0.058
Cutpoints					
τ_1	−0.432	1.187		−0.242	0.933
τ_2	−5.113	1.420		−4.120	1.066
Model Fit					
R^2_M		0.095		0.096	
R^2_O		0.129		0.128	
R^2_N		0.100		0.098	

Note: The results from significance tests are not reported for the cutpoints. SE = standard error, OR = odds ratio. X-standardized odds ratios are reported for continuous variables (age, income, education, ideology, and racial attitude), and un-standardized odds ratios are reported for binary variables (female and party identification). $N = 392$.

[a] Reference = Republican.

#$p < 0.10$, *$p < 0.05$, **$p < 0.01$, ***$p < 0.001$.

The coefficients for ideology and party identification vary across equations as well. Conservative ideology is associated with greater levels of opposition to welfare spending in both equations, but the effect is considerably stronger for mixed support versus support. We see the opposite pattern for party identification. Applying the parallel regression assumption would have the most severe consequences for education. Based on the results in Table 3.7, one would conclude that education is a significant predictor of welfare attitudes, but applying the parallel assumption averages a significant, negative coefficient with a positive coefficient to produce a nonsignificant, negative coefficient. Thus, if one were to use traditional ordinal methods, such as cumulative odds or adjacent category logit with proportional odds, then one would conclude that education is statistically unrelated to welfare spending attitudes. This is an example where a partial ordered regression model is clearly preferable over traditional models that require unrealistic assumptions. The pseudo R^2 measures range from 0.10 to 0.13 and the model explains more ordinal variation than nominal variation.

In Table 3.8, we present the results from constrained partial adjacent category models of welfare spending attitudes. The coefficients for the two party identification variables are

TABLE 3.8

Results from Constrained Partial Adjacent Category Ordered Regression Models of Welfare Spending Attitudes

	(Welfare Spending Attitude: 1 = Too Much, 2 = About Right, 3 = Too Little)				
	Logit			Probit	
	Coef.	SE	OR	Coef.	SE
Variables					
Age					
1 vs. 2	0.016*	0.007	0.762	0.012*	0.006
2 vs. 3	−0.004	0.009	1.069	−0.003	0.006
Female	−0.017	0.151	1.017	0.004	0.114
Income	−0.237***	0.070	1.311	−0.186***	0.053
Education					
1 vs. 2	0.075	0.046	0.812	0.059	0.037
2 vs. 3	−0.135*	0.058	1.454	−0.105*	0.043
Ideology					
1 vs. 2	−0.012	0.096	1.018	−0.020	0.075
2 vs. 3	−0.219#	0.124	1.373	−0.153#	0.084
Racial attitude	−0.239**	0.077	1.273	−0.180**	0.058
Variables with Partial Proportionality Constraints (PPC)					
Party identification[a]					
Democrat	1.244***	0.355	0.288	0.979***	0.236
Independent	1.141***	0.315	0.320	0.896***	0.214
PPC					
φ_2	0.345	0.484		0.310	0.457
Cutpoints					
τ_1	−0.429	1.187		−0.317	0.899
τ_2	−5.125	1.415		−3.977	0.951
Model Fit					
R^2_M		0.095			0.096
R^2_O		0.129			0.127
R^2_N		0.100			0.097

Note: SE = standard error, OR = odds ratio. X-standardized odds ratios are reported for continuous variables (age, income, education, ideology, and racial attitude), and un-standardized odds ratios are reported for binary variables (female and party identification). $N = 392$.

[a] Reference = Republican.

#$p < 0.10$, *$p < 0.05$, **$p < 0.01$, ***$p < 0.001$.

constrained to vary by a common factor between the first and second cutpoint equations. We assume that $\varphi_1 = 1$ and $\varphi_3 = 0$ in order to identify the models. This only saves one parameter, but it may be a useful constraint because the coefficients for Democrat and Independent display very similar patterns in the unconstrained partial models. The coefficients are 65% weaker in the second cutpoint equation in the logit model and 69% weaker in the probit model. The results for the remaining variables are very similar to the unconstrained partial model results.

In order to further examine the substantive significance of these relationships, we present AMEs in Table 3.9 based on the partial adjacent category logit models with ADCs in parentheses. Most of the independent variables have stronger associations with opposition to welfare spending than with support for it. For example, the average effect of identifying

TABLE 3.9

AMEs from Unconstrained and Constrained Partial Adjacent Category Logit Models of Welfare Spending Attitudes

	Too Much	About Right	Too Little
Unconstrained Partial Model			
Age	−0.003*	0.003[#]	0.001
	(−0.054)	(0.045)	(0.009)
Female	0.005	−0.001	−0.003
Income	0.067***	−0.020**	−0.046***
	(0.075)	(−0.027)	(−0.048)
Education	−0.007	0.020*	−0.012[#]
	(−0.023)	(0.055)	(−0.033)
Ideology	0.017	0.012	−0.029*
	(0.023)	(0.016)	(−0.039)
Party identification[a]			
Democrat	−0.290***	0.151[#]	0.139[#]
Independent	−0.259***	0.135*	0.125*
Racial attitude	0.067**	−0.020**	−0.047**
	(0.067)	(−0.024)	(−0.043)
Constrained Partial Model			
Age	−0.003*	0.003[#]	0.001
	(−0.054)	(0.045)	(0.009)
Female	0.005	−0.001	−0.003
Income	0.067***	−0.020**	−0.046***
	(0.075)	(−0.027)	(−0.048)
Education	−0.007	0.020*	−0.012[#]
	(−0.023)	(0.056)	(−0.033)
Ideology	0.017	0.012	−0.029*
	(0.023)	(0.016)	(−0.039)
Party identification[a]			
Democrat	−0.289***	0.148***	0.141***
Independent	−0.260***	0.137***	0.123***
Racial attitude	0.067**	−0.020**	−0.047**
	(0.067)	(−0.024)	(−0.043)

Note: The numbers in parentheses are average standard deviation discrete changes.

[a] Reference = Republican.

[#]$p < 0.10$, *$p < 0.05$, **$p < 0.01$, ***$p < 0.001$.

as a Democrat versus Republican is a 0.29 decrease in the predicted probability of welfare opposition, compared to a 0.14 increase in the predicted probability of welfare support. The pattern is very similar for age, although age differences in welfare attitudes are weaker. Negative racial attitudes are significantly associated with all three views of welfare spending, but the average effect is also stronger for opposition than support.

Ideology and education do not follow the same pattern, however. Political conservatism is associated with a decrease in support for welfare spending, but it is not significantly associated with opposition or mixed support. Education is the only variable in the model that has a stronger association with mixed support than the two extreme positions. A standard deviation increase in education is associated with an average increase of 0.055 in the predicted probability of mixed support. Education is associated with a decrease in the predicted probabilities for opposition and support, but the effects are either marginally significant ($p < 0.10$) or nonsignificant. The coefficients for education and ideology are allowed to freely vary across cutpoint equations in the partial adjacent category models in this example. Applying the parallel regression assumption for these variables would result in dramatically different conclusions. The AMEs for education and ideology are nonsignificant for all three outcome categories in the parallel adjacent category models.

The use of partial adjacent category models allows us to uncover a more nuanced relationship between education and welfare attitudes than one would find using linear regression or ordered regression models relying on the parallel regression assumption. We examine this relationship in greater detail by plotting the predicted probabilities for each category by the level of education (see Figure 3.3). Across the range of education, the

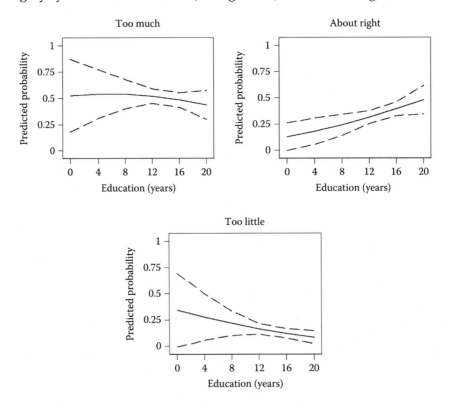

FIGURE 3.3
The effect of education on welfare spending attitudes from an unconstrained partial adjacent category logit model.

predicted probability of mixed support increases by 0.35 (from 0.13 to 0.48). There is also a substantial decrease in the predicted probability of support for welfare spending (0.34 to 0.09) and a more modest decrease in opposition (0.53 to 0.43). The confidence intervals are relatively wide between 0 and 10 years of schooling given that 87% of respondents have at least a high school education.

On balance, the results in this example suggest that education is associated with welfare attitudes in the United States, but it does not follow an ordinal pattern that traditional models such as the parallel adjacent category model would detect. The flexibility of partial models to allow for ordinal and nonordinal relationships in the same model enables us to uncover this asymmetrical effect of education. Researchers can also use nonparallel models for this purpose, which we introduce in the next chapter.

3.5 Dimensionality in Partial Models

The unconstrained and constrained partial models differ in terms of the manner in which they relax the parallel regression assumption, but these models actually are more closely related than the previous discussion may suggest. The concept of dimensionality links these models together (Fullerton and Xu 2015). According to Anderson (1984, p. 2), dimensionality "is determined by the number of linear functions required to describe the relationship." One may be able to think of two or more latent, continuous variables underlying a single ordinal outcome (Long 1997, p. 115). Each dimension could represent a different latent variable. Long and Freese (2014) use party identification as an example. The first dimension is the left/right continuum (Democrat $\rightarrow$ Republican), and the intensity of this affiliation represents a second dimension (weak $\rightarrow$ strong).

Anderson (1984) also developed a multidimensional version of the stereotype logit model. The maximum number of dimensions (D) is the minimum of the number of independent variables and the number of cutpoint equations ($M - 1$) (Anderson 1984, p. 5). The equation for the log odds in the multidimensional stereotype logit model is (Fullerton and Xu 2015)

$$\ln\left[\frac{\Pr(y = m \mid \mathbf{x})}{\Pr(y = M \mid \mathbf{x})}\right] = \tau_m - \sum_{q=1}^{d} \varphi_m^q \mathbf{x}\boldsymbol{\beta}^q \qquad (1 \le d \le D) \tag{3.8}$$

where d is the number of dimensions. In the partial adjacent category example, the maximum number of dimensions is 2 because there are only two cutpoint equations (1 vs. 2 and 2 vs. 3). The equation for the log odds in the two-dimensional stereotype logit model is

$$\ln\left[\frac{\Pr(y = m \mid \mathbf{x})}{\Pr(y = M \mid \mathbf{x})}\right] = \tau_m - \varphi_m^{(1)} \mathbf{x}\boldsymbol{\beta}^{(1)} - \varphi_m^{(2)} \mathbf{x}\boldsymbol{\beta}^{(2)} \tag{3.9}$$

Due to the additional constraints on φ necessary to identify the model, a stereotype logit model with the maximum number of dimensions is equivalent to the multinomial logit model (Anderson 1984; Long 1997; Agresti 2010).

The implication of this for partial ordered regression models is that the unconstrained partial model is equivalent to a constrained partial model with the maximum number of dimensions. Fullerton and Xu (2015) discuss this in the context of adjacent category models, but it applies to partial cumulative and continuation ratio models as well. We can use this connection between unconstrained and constrained partial models to develop a generalized partial model. For the adjacent class of models, the equation for the nonlinear adjacent probability is (Fullerton and Xu 2015)

$$\Pr\left(y = m \mid y = m \text{ or } y = m+1, \mathbf{x}, \boldsymbol{\omega}, \boldsymbol{\gamma}\right) = F\left(\tau_m - \mathbf{x}\boldsymbol{\beta} - \omega\eta_m - \sum_{q=1}^{d} \varphi_m^q \boldsymbol{\gamma} \lambda^q\right) \qquad (1 \le m < M) \quad (3.10)$$

where d represents the number of dimensions. A generalized partial model with the maximum number of dimensions (D) is equivalent to the unconstrained partial model, whereas a generalized partial model with a single dimension is equivalent to the constrained partial model. It is also possible to have generalized models with an intermediate number of dimensions (e.g., three of four possible dimensions), but it is not likely that one would be able to justify such a complex modeling strategy on theoretical grounds. The unconstrained partial models are perhaps the easiest to justify because they do not place additional constraints on the manner in which coefficients vary across cutpoint equations.

One may be inclined to compare the constrained and unconstrained partial models using nested model tests such as the likelihood ratio test. This is not appropriate, however, because φ is not identified if the slopes are constrained to equal zero under the null hypothesis (Anderson 1984, p. 12). As a result, researchers should compare the relative model fit of constrained and unconstrained partial models using alternative measures such as the Akaike information criterion and Bayesian information criterion or nonnested tests such as the Vuong test (Vuong 1989).

3.6 Conclusions

Partial models are flexible alternatives to parallel models that allow researchers to analyze ordinal outcomes within a regression framework. The key assumption of parallel regressions allows traditional ordinal models to be relatively parsimonious and to maintain a strict stochastic ordering in the relationships between independent and dependent variables. The parallel regression assumption constrains the coefficients for each independent variable to remain constant across cutpoint equations. For example, if a variable has an odds ratio of 1.5 in the first cutpoint equation, it will have an odds ratio of 1.5 in the remaining equations as well. If the coefficients vary substantially across equations for a particular variable, then the application of this constraint can lead to misleading results that affect the substantive conclusions. It is common for the parallel assumption to be a reasonable constraint for some variables in the model but not for others. Partial models are ideally suited in this case because they allow the researcher to relax the parallel assumption for a subset of the variables.

The most straightforward and perhaps intuitive way to relax the parallel regression assumption for a set of variables is to allow the coefficients to freely vary across the cutpoint equations. This is the approach that the unconstrained partial models take. Allowing the coefficients for one or more variables to freely vary across equations obviously adds complexity to the model because now there are $M-1$ slopes for every variable rather than just one slope. For instance, if there are five outcome categories and four cutpoint equations, then relaxing the parallel assumption in an unconstrained manner for four independent variables adds 12 parameters to the model (4 → 16 slopes). The constrained partial model also allows one to relax the parallel assumption for a subset of variables, but the coefficient variation across equations may be constrained for two or more variables. In this example, if we assume that two of these four variables have coefficients that vary across cutpoint equations by a common set of factors, then we would only need to add nine parameters (4 slopes → 10 slopes and 3 common factors). Constrained partial models may be particularly useful in small samples because they are more parsimonious than unconstrained partial models.

In the three empirical examples, the partial models revealed more nuanced relationships than we found in the parallel models. The partial cumulative models of self-rated health relaxed the parallel regression assumption for age, age-squared, employment, and marital status. Each of these variables tended to have stronger effects on poor or fair health than good or excellent health. The parallel restriction resulted in a nonsignificant coefficient for marital status in the parallel cumulative models. However, in the partial models, we saw that being married significantly reduces the odds of poor health but has little to no effect on good or excellent health.

Similarly, the partial continuation ratio models of educational attainment revealed that the effects of mother's and father's education are weaker in later educational stages, which is consistent with previous research. It is important to keep in mind that this may be due to changing effects, differences in residual variance, or sample selection bias (Mare 2006; Buis 2011). If sensitivity analyses suggest that selection bias is a serious concern, then one should consider an alternative method such as probit with sample selection (Heckman 1979; Holm and Jæger 2011). If it is an issue of changes in residual variance (i.e., scaling), then the heterogeneous choice model is worthy of consideration (Williams 2009, 2010).

The results from the partial adjacent category models also forced us to reconsider several conclusions regarding Whites' welfare attitudes. According to the parallel models, education and political ideology are not associated with attitudes toward welfare spending. However, the partial models indicate that education significantly increases the probability of mixed support, and conservative ideology significantly decreases the probability of support for welfare spending. By relaxing the parallel regression assumption, we see that education and ideology have stronger effects on welfare support than opposition compared to mixed support. The parallel adjacent category model averages these two coefficients, which results in nonsignificant effects for education and ideology.

Finally, it is possible that the choice of a particular approach (cumulative, stage, or adjacent) may affect the substantive findings. For the three examples in this chapter, there are noticeable differences in the raw coefficients across the three approaches for the subset of variables without the parallel regression assumption. This is due to the fact that the cutpoint equations are based on different subsets of categories, and the coefficients for these variables are allowed to vary across equations. For example, in an adjacent category model of self-rated health, the third cutpoint equation compares good versus excellent health, whereas the cumulative model compares poor, fair, or good versus excellent health. It is reasonable to expect that the coefficients in these equations would be somewhat different. However, in our three examples, there were only modest differences in the predicted probabilities, which means that the substantive conclusions were not affected by the decision to use one approach rather than another.

The choice among cumulative, stage, and adjacent models may depend on interpretation preferences, desired comparisons in the cutpoint equations, and assumptions regarding the data-generating process. For example, one should choose a cumulative or adjacent model rather than a stage model to analyze subjective social class because the assumption of a sequential decision-making process is not applicable. One does not have to identify with the "lower" or "working" classes before identifying with the "middle" or "upper middle" classes.

3.6.1 Guidelines for Choosing a Partial or Parallel Ordered Regression Model

Researchers should consider two main factors when deciding whether to use a partial or parallel ordered regression model to analyze an ordinal outcome. First, is there any ambiguity regarding the ordering of outcome categories, and is the category order important for the statistical relationships? If there is no ambiguity and the category order is a salient characteristic, then parallel models are strongly preferred because the parallel regression assumption guarantees the ordinal nature of the relationships. However, midpoint categories such as "neither agree nor disagree" may introduce enough ambiguity to warrant consideration of alternative models. This category could represent a view located between agreement and disagreement, or it could be a form of nonresponse (Sturgis et al. 2014). If there is a substantial amount of ambiguity, then one should not use a parallel or partial model. In this case, the nonparallel adjacent category logit model is preferable because it is a re-parameterization of the multinomial logit model for nominal outcomes.

Second, is the parallel regression assumption reasonable for every independent variable? Determining the reasonableness of the parallel regression assumption requires the consideration of several alternative models and should not be based on a single test. If the use of the parallel regression assumption does not lead to different substantive conclusions, then the assumption is reasonable and the parallel model is preferred because it is more parsimonious. We discuss several formal and informal tests of the parallel regression assumption in Chapter 5.

Nonparallel ordered models are another alternative to consider when the parallel regression assumption is too restrictive for one or more independent variables in the model. The most widely used nonparallel model, multinomial logit, is actually a method designed for nominal outcomes. The nonparallel models from the cumulative and stage approaches have also gained popularity in recent years (generalized ordered logit and sequential logit). These models relax the parallel assumption for every independent variable. We will consider these models in the next chapter.

3.7 Appendix

3.7.1 Stata Codes for Partial Ordered Logit Models

Unconstrained Partial Cumulative Logit Model of Self-Rated Health

```
gologit2 ghealth age agesq female black hisp2 othrace educ employed
   lninc married, npl(age agesq employed married) link(logit)
```

Constrained Partial Cumulative Logit Model of Self-Rated Health

```
constraint 1 [eq2]age=[eq1]age*.3754905
constraint 2 [eq2]agesq=[eq1]agesq*.3754905
constraint 3 [eq3]age=[eq1]age*.3480221
constraint 4 [eq3]agesq=[eq1]agesq*.3480221
gologit2 ghealth age agesq female black hisp2 othrace educ employed
  lninc married, npl(age agesq employed married) constraint(1-4)
```

Unconstrained Partial Continuation Ratio Logit Model of Educational Attainment

```
constraint 1  [#1]1.female=[#2]1.female
constraint 2  [#1]1.female=[#3]1.female
constraint 3  [#1]1.female=[#4]1.female
constraint 4  [#1]1.black=[#2]1.black
constraint 5  [#1]1.black=[#3]1.black
constraint 6  [#1]1.black=[#4]1.black
constraint 7  [#1]1.hisp2=[#2]1.hisp2
constraint 8  [#1]1.hisp2=[#3]1.hisp2
constraint 9  [#1]1.hisp2=[#4]1.hisp2
constraint 10 [#1]1.othrace=[#2]1.othrace
constraint 11 [#1]1.othrace=[#3]1.othrace
constraint 12 [#1]1.othrace=[#4]1.othrace
seqlogit degree5 age i.female i.black i.hisp2 i.othrace maeduc paeduc,
  tree(1:2 3 4 5 , 2:3 4 5,3:4 5,4:5) constraint(1/12)
```

Constrained Partial Continuation Ratio Logit Model of Educational Attainment

```
constraint 13 [#2]maeduc=[#1]maeduc*.6712194
constraint 14 [#3]maeduc=[#1]maeduc*.5181975
constraint 15 [#4]maeduc=[#1]maeduc*-.0281166
constraint 16 [#2]paeduc=[#1]paeduc*.6712194
constraint 17 [#3]paeduc=[#1]paeduc*.5181975
constraint 18 [#4]paeduc=[#1]paeduc*-.0281166
seqlogit degree5 age i.female i.black i.hisp2 i.othrace maeduc paeduc,
  tree(1:2 3 4 5,2:3 4 5,3:45,4:5) constraint(1/18)
```

Unconstrained Partial Adjacent Category Logit Model of Welfare Attitudes

```
constraint 1 [too_little]1.female=[too_much]1.female*-1
constraint 2 [too_little]lninc=[too_much]lninc*-1
constraint 3 [too_little]workblks=[too_much]workblks*-1
mlogit natfare2 age i.female educ lninc polviews i.democrat i.indep
  workblks, base(2) constraint(1/3)
```

Constrained Partial Adjacent Category Logit Model of Welfare Attitudes

```
constraint 4 [too_little]1.democrat=[too_much]1.democrat*-.3445967
constraint 5 [too_little]1.indep= [too_much]1.indep*-.3445967
mlogit natfare2 age i.female educ lninc polviews i.democrat i.indep
  workblks, base(2) constraint(1/5)
```

3.7.2 R Codes for Partial Ordered Logit Models

Unconstrained Partial Cumulative Logit Model of Self-Rated Health

```
ppoLogit <- vglm(as.ordered(ghealth) ~ age + agesq + # bs(age)
          female
          + black + hisp2 + othrace
          + educ + employed + lninc + married,
          cumulative(link="logit",
                parallel = TRUE ~ female + black + hisp2
                              + othrace + educ + lninc
#               + age + agesq + employed + married # zz-1,
                reverse = TRUE),
        trace = TRUE, # Rank = 1,
#        etastart = predict(poLogit),
#         Index.Corner = 2,
#        stepsize = 0.15,
#        maxit = 2,
          data = mydta)
summary(ppoLogit)
```

Constrained Partial Cumulative Logit Model of Self-Rated Health

```
ppcLogit <- rrvglm(as.ordered(ghealth) ~ age + agesq + female
            + black + hisp2 + othrace
            + educ + employed + lninc + married,
            cumulative(link="logit",
            parallel = TRUE ~ female + black + hisp2 +
                        othrace + educ + lninc - 1,
            reverse = TRUE),
            # Index.Corner = 3,
            trace = TRUE, Rank = 1,
            noRRR = ~ female + black + hisp2 + othrace + educ + lninc +
              married + employed,
            data = mydta)
summary(ppcLogit)
```

Unconstrained Partial Continuation Ratio Logit Model of Educational Attainment

```
crPpoLogit <- vglm(as.ordered(degree5) ~ age + female + black
            + hisp2 + othrace + maeduc + paeduc,
            cratio(link="logit",
                parallel = TRUE ~ female + black + hisp2 + othrace -1,
                reverse = FALSE),
            trace = TRUE, # Rank = 1,
            data = mydta)
summary(crPpoLogit)
```

Constrained Partial Continuation Ratio Logit Model of Educational Attainment

```
crPpcLogit <- rrvglm(as.ordered(degree5) ~ age + female + black
            + hisp2 + othrace + maeduc + paeduc,
            cratio(link="logit",
            parallel = TRUE ~ female + black + hisp2 + othrace - 1,
```

```
                    reverse = FALSE),
                    # Index.Corner = 3,
                    trace = TRUE, Rank = 1,
                    noRRR = ~ age + female + black + hisp2 + othrace,
                    data = mydta)
summary(crPpcLogit)
```

Unconstrained Partial Adjacent Category Logit Model of Welfare Attitudes

```
acPpoLogit <- vglm(as.ordered(natfare2) ~ age + female + educ
              + lninc + polviews + democrat
              + indep + workblks,
              acat(link="loge",
                parallel = TRUE ~ female + lninc + workblks -1 ,
                reverse = FALSE),
              trace = TRUE, # Rank = 1,
              data = mydta)
summary(acPpoLogit)
```

Constrained Partial Adjacent Category Logit Model of Welfare Attitudes

```
acPpcLogit <- rrvglm(as.ordered(natfare2) ~ age + female + educ
              + lninc + polviews + democrat
              + indep + workblks,
              acat(link="loge",
              parallel = TRUE ~ female + lninc + workblks - 1,
              reverse = FALSE),
              # Index.Corner = 3,
              trace = TRUE, Rank = 1,
              noRRR = ~ age + female + educ + lninc + polviews + workblks,
              data = mydta)
summary(acPpcLogit)
```

4

Nonparallel Models

Nonparallel models are ordered regression models that relax the parallel regression assumption for every independent variable in the model. The slopes for every variable are allowed to freely vary across the binary cutpoint equations. Nonparallel models are not ordinal models in the strictest sense because they do not impose constraints to maintain a strict stochastic ordering. However, ordinal patterns may emerge without the use of the parallel constraints. The nonparallel cumulative and continuation ratio models are ordinal in a weaker sense because reordering the outcome categories (e.g., 1,2,3,4 → 3,1,4,2) affects the coefficients and significance tests. In other words, the models are not "permutation invariant" (McCullagh 1980, p. 116). The nonparallel adjacent category model, however, does display permutation invariance. It is equivalent to multinomial logit, which is a model designed for nominal outcomes. Therefore, it is not an ordinal method even in the weaker sense.

The three classes of nonparallel models (cumulative, stage, and adjacent) are flexible alternatives to the parallel and partial ordered regression models, but this flexibility comes at a price. Nonparallel models are also the most inefficient models that we consider in this book in terms of the number of parameters. They require $M - 1$ coefficients for each independent variable, where M represents the number of response categories. By contrast, the parallel models only require one coefficient per variable in addition to the cutpoints. Relaxing the parallel regression assumption for the entire model can add a substantial number of parameters if there are a lot of response categories and independent variables. For example, if $M = 5$ and there are 10 independent variables, then the nonparallel model requires 30 additional coefficients (40 vs. 10 slopes). As a result, a large sample size is necessary in order to ensure an adequate number of cases per parameter (see Long 1997, pp. 53–54).

The nonparallel adjacent category model is more widely used than the cumulative or stage models because it is simply a reparameterization of the multinomial logit model. It may also be the preferred method for ordinal outcomes if there is some ambiguity in the category order and the effects of several variables are noticeably different across equations. The nonparallel continuation ratio model is frequently used to study sequential processes, such as educational attainment, and is also known as "sequential logit/ probit" or the "Mare model" in the context of educational research (Mare 1980, 1981). The nonparallel cumulative model has also gained popularity in recent years with software advancements that have allowed researchers to easily estimate this model in programs such as R (Yee 2010), SAS (Lall et al. 2002), and Stata (Williams 2006).* Nonparallel models may provide a better fit to the data than parallel models for several reasons, including model mis-specification, asymmetrical effects, or heteroscedastic errors (Brant 1990; Williams 2010, 2015). The choice of one nonparallel model over another depends largely on the substantive importance of the response categories and whether or not the outcome is the result of a sequential decision-making process.

---

* Yee's (2010) R software package, VGAM, can estimate a broad class of models known as "vector generalized linear models" (Yee 2015), which encompass the parallel, partial, and nonparallel ordered regression models.

4.1 The Nonparallel Cumulative Model

In Chapter 3, we discussed two problems with the partial cumulative model: negative predicted probabilities and the lack of a latent variable motivation. These problems apply to the nonparallel cumulative model as well. The potential for negative predicted probabilities is even greater in the nonparallel model because relaxing the parallel assumption for every variable increases the chances that the cumulative probability curves will intersect within the observed range of values. If the lines do intersect, then the cumulative probability for a category, m, will be greater than the cumulative probability for the next highest category, $m + 1$. For example, if the cumulative probability for category 3 is 0.50 and the cumulative probability for category 4 is 0.49, then the probability for category 4 must be −0.01. The nonparallel cumulative model is a useful model despite this internal inconsistency if there are no observations with negative predicted probabilities.

Researchers need to be aware of this potential problem and examine the observed range of predicted probabilities for each model. In some cases, the presence of negative predicted probabilities may pose convergence problems for the model.* If the model does not converge, then one could specify starting values based on the results from a parallel model or choose to relax the parallel regression assumption for only a few variables in the model. In addition, the predicted probabilities will be nonnegative if one either retains the parallel assumption for every continuous predictor or recodes them into binary variables. A model with only binary independent variables will not have negative predicted probabilities because the cumulative probability curves will intersect outside the observed data range (Hedeker et al. 1999, p. 64). Finally, McKinley et al. (2015) propose using a Bayesian framework to address this problem.

Relaxing the parallel regression assumption also means that we must treat the outcome as a set of discrete categories that do not correspond to different points along an underlying, latent continuous variable (Greene and Hensher 2010, pp. 190–192). However, the latent variable motivation is not necessary for estimation or interpretation (McCullagh 1980, p. 110). Rather than assuming the ordinal response variable reflects an underlying continuous distribution, one must treat it as a series of binary outcomes. For example, a four-category Likert scale question (1 = strongly agree, 2 = agree, 3 = disagree, and 4 = strongly disagree) corresponds to three binary outcomes in the cutpoint equations: 1 versus 2–4, 1–2 versus 3–4, and 1–3 versus 4. These three binary outcomes represent strong agreement, agreement, and lack of strong disagreement. We must assume that these represent three distinct outcomes rather than different points in the same distribution.

The nonparallel cumulative model is a nonlinear cumulative probability model that relaxes the parallel regression assumption for every variable. The general form of the equation for the nonlinear cumulative probability model is

$$\Pr(y \leq m \mid \mathbf{x}) = F(\tau_m - \mathbf{x}\boldsymbol{\beta}_m) \qquad (1 \leq m < M) \tag{4.1}$$

---

* The gologit2 program in Stata (Williams 2006) is generally able to converge and provides warnings when there are observations with negative predicted probabilities. However, other programs may not provide such warnings regardless of whether or not the model converges. Therefore, we recommend examining the range of predicted probabilities in order to determine if the internal inconsistency in the nonparallel cumulative model is of concern for that particular model.

where F typically represents the cumulative distribution function (CDF) for the logit, probit, or complementary log log model, τ_m are cutpoints, $\mathbf{x}$ is a vector of independent variables, and $\boldsymbol{\beta}_m$ is a vector of coefficients that freely vary across cutpoint equations.

The original, ordinal outcome is recoded into a series of $M - 1$ binary outcomes corresponding to each cutpoint equation. If $M = 4$, then the nonparallel cumulative model estimates the following three binary cutpoint equations simultaneously based on different points in the cumulative distribution:

$$\text{Cutpoint equation \#1: } \Pr(y{\le}1|\mathbf{x}) = F(\tau_1 - \mathbf{x}\boldsymbol{\beta}_1)$$
$$\text{Cutpoint equation \#2: } \Pr(y{\le}2|\mathbf{x}) = F(\tau_2 - \mathbf{x}\boldsymbol{\beta}_2) \quad\quad (4.2)$$
$$\text{Cutpoint equation \#3: } \Pr(y{\le}3|\mathbf{x}) = F(\tau_3 - \mathbf{x}\boldsymbol{\beta}_3)$$

The values for the cutpoints (τ_m) and the coefficients for the independent variables ($\boldsymbol{\beta}_m$) vary across equations, whereas the values for the independent variables ($\mathbf{x}$) remain constant across equations. The nonparallel cumulative probit model is equivalent to Terza's (1985) generalization of the ordinal probit model, which allows for individual heterogeneity in the cutpoints by interacting them with the independent variables (see Greene and Hensher 2010, pp. 209–214).

4.1.1 Example: Nonparallel Cumulative Models of Self-Rated Health

We will illustrate the nonparallel cumulative model using the general self-rated health example from the 2010 GSS that we introduced in Chapter 1. We present coefficients and odds/hazard ratios from the nonparallel cumulative models using the logit, probit, and complementary log log links in Table 4.1. The odds and hazard ratios are x-standardized for the continuous variables (age, age-squared, education, and log income). All of the observations in our sample have probabilities that fall within the 0–1 range using the logit and probit links. However, there is one case with a negative predicted probability using the complementary log log link. Given that the predicted probabilities fall within the 0–1 range for two of the three link functions, this internal inconsistency does not appear to be a serious problem for this example.[*]

There are three sets of coefficients for each variable because the parallel regression assumption is relaxed for the entire model. Based on the raw coefficients and odds/hazard ratios, several variables (age, age-squared, employed, and married) have stronger effects in the first cutpoint equation than in the next two equations. This indicates that they have the strongest effects on poor health. For example, being employed is associated with a decrease in the odds of poor health by a factor of 0.20 or 80%.[†] By contrast, employment is only associated with a decrease in the odds of poor to fair health by a factor of 0.45 or 55%. The odds ratio is considerably closer to one for poor to good health and is not statistically significant. These results suggest that employed workers are less likely to be in poor health, but they are not more likely to be in excellent health than individuals that are unemployed or not in the labor force. There is a similar pattern for marital status. Being married is associated with a lower odds or hazard rate for poor health, but it is not associated with a higher odds

---

[*] The case with a negative predicted probability for "fair" health has an extreme value (minimum) for the log of income. Recoding income into a binary variable is one solution to this problem.

[†] For all of the interpretations of odds/hazard ratios, we assume that the remaining variables in the model are held constant.

Ordered Regression Models

TABLE 4.1

Results from Nonparallel Cumulative Ordered Regression Models of Self-Rated Health

| | (General Self-Rated Health: 1 = Poor, 2 = Fair, 3 = Good, 4 = Excellent) | | | | | | | |
| | Logit | | | Probit | | Complementary Log Log | | |
	Coef.	SE	OR	Coef.	SE	Coef.	SE	HR
Variables								
Age								
1 vs. 2–4	−0.253***	0.061	65.704	−0.116***	0.028	−0.246***	0.058	58.707
1–2 vs. 3–4	−0.087***	0.025	4.207	−0.051***	0.015	−0.073***	0.020	3.327
1–3 vs. 4	−0.064*	0.027	2.877	−0.038*	0.015	−0.035*	0.014	1.771
Age-squared[a]								
1 vs. 2–4	0.214***	0.056	0.027	0.098***	0.025	0.209***	0.053	0.030
1–2 vs. 3–4	0.070**	0.024	0.307	0.041**	0.015	0.059**	0.020	0.372
1–3 vs. 4	0.046#	0.027	0.466	0.027#	0.015	0.026#	0.013	0.650
Female								
1 vs. 2–4	0.161	0.274	0.852	0.009	0.139	0.147	0.251	0.863
1–2 vs. 3–4	−0.096	0.144	1.101	−0.065	0.085	−0.126	0.117	1.134
1–3 vs. 4	0.075	0.145	0.928	0.033	0.085	0.016	0.078	0.984
Race/ethnicity[b]								
Black								
1 vs. 2–4	−0.220	0.350	1.247	−0.172	0.179	−0.134	0.318	1.143
1–2 vs. 3–4	0.225	0.216	0.799	0.111	0.126	0.229	0.174	0.796
1–3 vs. 4	0.028	0.227	0.972	0.024	0.132	0.056	0.121	0.946
Latino								
1 vs. 2–4	−0.153	0.450	1.165	−0.167	0.222	−0.138	0.413	1.148
1–2 vs. 3–4	−0.409#	0.224	1.505	−0.262#	0.134	−0.341*	0.171	1.406
1–3 vs. 4	−0.378	0.270	1.459	−0.207	0.152	−0.150	0.131	1.162
Other race/ethnicity								
1 vs. 2–4	0.056	1.041	0.945	0.096	0.497	0.097	1.008	0.908
1–2 vs. 3–4	−0.472	0.353	1.603	−0.292	0.203	−0.282	0.282	1.326
1–3 vs. 4	−0.589	0.379	1.801	−0.335	0.216	−0.256	0.192	1.291
Education								
1 vs. 2–4	0.054	0.050	0.846	0.020	0.026	0.053	0.044	0.849
1–2 vs. 3–4	0.103***	0.027	0.729	0.060***	0.016	0.083***	0.020	0.775
1–3 vs. 4	0.117***	0.028	0.699	0.067***	0.016	0.058***	0.014	0.838
Employed								
1 vs. 2–4	1.625***	0.368	0.197	0.731***	0.164	1.575***	0.356	0.207
1–2 vs. 3–4	0.795***	0.154	0.452	0.460***	0.092	0.675***	0.126	0.509
1–3 vs. 4	0.268	0.164	0.765	0.155	0.095	0.146#	0.086	0.864
Log income								
1 vs. 2–4	0.342*	0.143	0.666	0.151*	0.073	0.348**	0.129	0.661
1–2 vs. 3–4	0.298***	0.075	0.701	0.173***	0.045	0.229***	0.055	0.762
1–3 vs. 4	0.287***	0.084	0.711	0.160***	0.046	0.136***	0.039	0.851
Married								
1 vs. 2–4	0.665*	0.326	0.514	0.361*	0.157	0.607*	0.309	0.545
1–2 vs. 3–4	0.249	0.159	0.779	0.159#	0.093	0.203	0.130	0.816
1–3 vs. 4	0.047	0.163	0.954	0.039	0.095	0.037	0.085	0.963

(Continued)

TABLE 4.1 *(Continued)*

Results from Nonparallel Cumulative Ordered Regression Models of Self-Rated Health

	(General Self-Rated Health: 1 = Poor, 2 = Fair, 3 = Good, 4 = Excellent)							
	Logit			Probit		Complementary Log Log		
	Coef.	SE	OR	Coef.	SE	Coef.	SE	HR
Cutpoints								
τ_1	−4.777	2.010		−2.609	0.938	−4.622	1.889	
τ_2	1.457	0.808		0.804	0.476	0.600	0.633	
τ_3	3.930	0.860		2.218	0.494	1.582	0.443	
Model fit								
R^2_M		0.086			0.083		0.085	
R^2_O		0.108			0.102		0.105	
R^2_N		0.060			0.055		0.058	

Note: The results from significance tests are not reported for the cutpoints. SE = standard error, OR = odds ratio, HR = hazard ratio. X-standardized odds and hazard ratios are reported for continuous variables (age, age-squared, education, and log income), and un-standardized odds and hazard ratios are reported for binary variables (female, race/ethnicity, employed, and married). N = 1129.

[a] Coefficients and standard errors are multiplied by 100 for ease of presentation.

[b] Reference = White.

#$p < 0.10$, *$p < 0.05$, **$p < 0.01$, ***$p < 0.001$.

or hazard rate for good or excellent health. Age is statistically significant in each cutpoint equation, but the coefficients are substantially larger in the first cutpoint equation.

We see the opposite pattern for the relationship between education and health. The coefficients for education are not significant in the first cutpoint equation. In the second and third equations, the coefficients are larger and statistically significant. A standard deviation increase in years of schooling is associated with a decrease in the odds of poor to good health by a factor of 0.70 or 30%. The inverse of the odds ratio (1/0.70) is 1.43, which means that the odds of excellent health increase by a factor of 1.43 or 43%. Therefore, the results indicate that education is not associated with the odds of poor health, but it is associated with a significant increase in the odds of excellent health.

The remaining variables in the models have coefficients that display less variation across the cutpoint equations. Based on these results, sex and race/ethnicity differences in self-rated health are not statistically significant after controlling for the other variables in the model. Although the coefficients vary somewhat across the equations and even change sign, they fail to reach statistical significance in all but one case. On balance, the parallel regression assumption appears to be reasonable for sex and race/ethnicity. A comparison of raw coefficients and odds/hazard ratios across cutpoint equations is just one of many informal tests of the parallel regression assumption, which we discuss in more detail in the next chapter. Finally, the model fit statistics suggest that the models explain more ordinal variation than nominal variation (10%–11% vs. 6%).

In order to examine the substantive effect size for each variable, we present average marginal effects based on the nonparallel cumulative logit model in Table 4.2. We also report average discrete changes for continuous variables in parentheses. Odds and hazard ratios can be misleading as measures of effect size because they do not necessarily correspond with large changes in the predicted probabilities. This appears to be the case to

TABLE 4.2

Average Marginal Effects from a Nonparallel Cumulative Logit Model of Self-Rated Health

	Poor	Fair	Good	Excellent
Age	0.012***	0.004	−0.004	−0.011*
	(0.531)	(−0.239)	(−0.151)	(−0.141)
Age-squared[a]	−0.010***	−0.002	0.004	0.008#
	(−0.053)	(−0.108)	(0.010)	(0.151)
Female	−0.007	0.024	−0.030	0.013
Race/ethnicity[b]				
Black	0.011	−0.049	0.033	0.005
Latino	0.007	0.069	−0.014	−0.061
Other race/ethnicity	−0.003	0.091	0.002	−0.090#
Education	−0.003	−0.016***	−0.002	0.020***
	(−0.007)	(−0.045)	(−0.014)	(0.067)
Employed	−0.070***	−0.078**	0.102**	0.046#
Log income	−0.016*	−0.037**	0.002	0.050***
	(−0.016)	(−0.042)	(−0.005)	(0.063)
Married	−0.029*	−0.015	0.036	0.008

Note: The numbers in parentheses are average standard deviation discrete changes.
[a] Coefficient multiplied by 100 for ease of presentation.
[b] Reference = White.
#$p < 0.10$, *$p < 0.05$, **$p < 0.01$, ***$p < 0.001$.

some extent for age and age-squared. On average, the effects of age and age-squared on the predicted probabilities for each level of health are more modest than the odds/hazard ratios would suggest.

The average effects of sex and race/ethnicity on the predicted probabilities are not statistically significant, which is what we would expect based on the raw coefficients and odds/hazard ratios. Additionally, marriage only has a significant effect on the probability of poor health, which is consistent with the previous results. On average, being married is associated with a 0.03 decrease in the predicted probability of poor health. Education, employment, and log income have more consistent effects across levels of health. Each of these variables is associated with an increase in the probability of good or excellent health and a decrease in the probability of poor or fair health. Employment is the most consistent across levels of health in terms of statistical significance. On average, employment is associated with decreases of 0.07 and 0.08 in the predicted probabilities of poor and fair health, respectively. Employment is also associated with an average increase of 0.10 in the predicted probability of good health. Its effect on excellent health is weaker and only marginally significant.

There is also a strong association between log income and self-rated health. We explore this relationship in more detail by plotting the predicted probabilities for each health category across the range of log income in Figure 4.1.* Income has the largest effects on the predicted probabilities of fair and excellent health. Across the range of log income,

---

* The predicted probabilities and 95% confidence intervals in this chapter are based on the models using the logit link with the remaining variables held constant at their means.

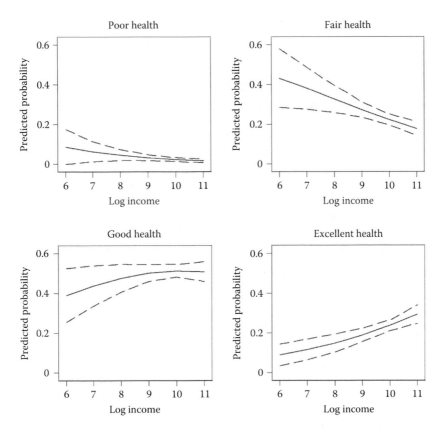

FIGURE 4.1
The effect of log income on general self-rated health from a nonparallel cumulative logit model.

the predicted probability of fair health decreases by 0.25 and the predicted probability of excellent health increases by 0.20. The predicted probability of good health also increases by 0.12 across the range of log income, whereas the predicted probability of poor health decreases by 0.07. The confidence bands are widest at the lowest income levels due to the smaller number of cases at the lowest income levels.

The results from this example highlight the usefulness of the nonparallel cumulative model. It is common for the coefficients from ordered regression models to vary at least to some degree across cutpoint equations. Traditional ordered regression models assume that the outcome is unidimensional and the effects of each variable are symmetrical. In other words, the effect of a variable at one extreme of the outcome distribution should be the mirror image of its effect at the other extreme. For example, we assume the associations of a variable such as income with poor and excellent health are equal in magnitude but in opposite directions. Nonparallel models allow variables to have asymmetrical effects, which means that the effect at one extreme is not simply the opposite of its effect at the other extreme. In the context of self-rated health, we observed several variables with noticeably different effects on the highest and lowest health categories. For instance, marriage is associated with a lower probability of poor health, but it has virtually no effect on fair to excellent health. As a result of this asymmetry, the traditional parallel cumulative model produces a nonsignificant raw coefficient for marriage because the significant effect on poor health is averaged together with the weaker, nonsignificant effects on the remaining health categories.

The main drawback in the nonparallel model is the relative lack of parsimony. In this example, relaxing the parallel regression assumption requires 20 additional slopes. The number of parameters in the model increases from 13 to 33. However, the added complexity is tolerable in this model due to the large sample size ($N = 1{,}129$). Additionally, we cannot refer to an underlying continuous health measure. Finally, negative predicted probabilities are possible in nonparallel cumulative models but not in nonparallel continuation ratio or adjacent models. There are no cases in our sample with negative predicted probabilities using the logit and probit links, but there is one case with a negative predicted probability in the complementary log log model. This particular link function is asymmetric, which means that reversing the outcome categories potentially changes the results. In this case, reverse-coding health solves the problem of negative predicted probabilities. Coding poor health as the highest or lowest category is arbitrary. Therefore, we could simply reverse the coding in order to ensure that the predicted probabilities are within the proper 0–1 range.

4.2 The Nonparallel Continuation Ratio Model

Continuation ratio or "stage" models are best suited for ordinal outcomes that represent an irreversible series of sequential decisions or stages (Fienberg 1980). Nonparallel continuation ratio models are also known as "sequential" models because of the nature of the data generating process (Amemiya 1981; Tutz 1991). The assumption of a sequential process is not appropriate for many ordinal dependent variables. Likert scale questions are good examples because there is no logical starting point and one does not have to (dis) agree with a statement before taking the opposite position. Self-rated health, which we examined in the previous section, is another good example because one does not have to begin with poor or excellent health and then transition to the other extreme. Examples of ordinal outcomes that are appropriate for sequential or continuation ratio models include educational attainment (Mare 1980, 1981), employment stability (Kahn and Morimune 1979), discussion networks (Liao 1994; Liao and Stevens 1994), and vote overreporting (Fullerton et al. 2007).

Negative predicted probabilities and the absence of a latent variable motivation are not a problem in the nonparallel continuation ratio model, but the conditional nature of the probability model poses a different set of potential problems. As we mentioned in the previous two chapters, sample selection bias in the later stages and differences in unobserved heterogeneity across stages make it difficult to establish "causal" effects (Cameron and Heckman 1998; Mare 2011). The coefficients from nonparallel continuation ratio models may vary across stages because of differences in the "true" effects, scaling effects due to changes in unobserved heterogeneity, or because of sample selection bias in later stages. As a result, we should view the raw coefficients in each cutpoint equation as descriptive rather than causal (Xie 2011), and we should use measures based on predicted probabilities rather than raw coefficients or odds ratios to make comparisons across equations (Buis 2011). However, probability measures are also potentially biased by the sample selection process through the progression of stages (Cameron and Heckman 1998). Nonparallel continuation ratio models are useful for modeling sequential processes, but researchers should be aware of these limitations as well.

The nonparallel continuation ratio model is a nonlinear conditional probability model that relaxes the parallel regression assumption for every variable. The general form of the equation for the nonlinear conditional probability model is

$$\Pr(y = m \mid y \geq m, \mathbf{x}) = F(\tau_m - \mathbf{x}\boldsymbol{\beta}_m) \qquad (1 \leq m < M) \tag{4.3}$$

where F typically represents the CDF for the logit, probit, or complementary log log model, τ_m are cutpoints, $\mathbf{x}$ is a vector of independent variables, and $\boldsymbol{\beta}_m$ is a vector of coefficients that freely vary across cutpoint equations.

The original, ordinal outcome is recoded into a series of $M - 1$ binary outcomes corresponding to each cutpoint equation. If $M = 4$, then the nonparallel continuation ratio model estimates the following three binary cutpoint equations simultaneously:

Cutpoint equation #1: $\Pr(y = 1 \mid y \geq 1, \mathbf{x}) = F(\tau_1 - \mathbf{x}\boldsymbol{\beta}_1)$

Cutpoint equation #2: $\Pr(y = 2 \mid y \geq 2, \mathbf{x}) = F(\tau_2 - \mathbf{x}\boldsymbol{\beta}_2)$

Cutpoint equation #3: $\Pr(y = 3 \mid y \geq 3, \mathbf{x}) = F(\tau_3 - \mathbf{x}\boldsymbol{\beta}_3)$

The sample gets progressively smaller in later stages because only the respondents that "survive" or make the transition are included in subsequent equations. The values for the cutpoints (τ_m) and the coefficients for the independent variables ($\boldsymbol{\beta}_m$) vary across equations, whereas the values for the independent variables ($\mathbf{x}$) remain constant across equations.

4.2.1 Example: Nonparallel Continuation Ratio Models of Educational Attainment

We will illustrate the nonparallel continuation ratio model using the educational attainment example from the 2010 GSS that we introduced in Chapter 1. We present coefficients and odds/hazard ratios from the nonparallel continuation ratio models using the logit, probit, and complementary log log links in Table 4.3. The odds and hazard ratios are x-standardized for the continuous variables (age, mother's education, and father's education). There are five educational categories and therefore four cutpoint equations or stages in the educational attainment process. The parallel regression assumption is relaxed for the entire model, which means the coefficients will vary across stages for every independent variable.

In the first stage, the outcome is a binary indicator for less than high school versus high school or more. Parent's education and race/ethnicity have statistically significant coefficients in this first stage, whereas age differences are nonsignificant and sex differences are only marginally significant. The Latino/White gap is the only significant racial or ethnic difference. Identifying as a Latino (vs. White) is associated with an increase in the odds of attaining less than a high school degree by a factor of 1.97 or 97%. Both measures of parent's education also have statistically significant coefficients. A standard deviation increase in mother's years of schooling is associated with a decrease in the hazard rate for attaining less than a high school degree by a factor of 0.52 or 48%. The hazard ratio for father's education indicates a somewhat weaker association.

In the second stage, the outcome is a binary indicator for a high school degree versus an associate's degree or more. In this stage, the sample is limited to respondents that successfully made the first transition. In other words, they achieved at least a high school degree. Sex is the only variable that does not have a significant coefficient in this equation. Age and parent's education are associated with decreases in the odds and hazard rates.

TABLE 4.3

Results from Nonparallel Continuation Ratio Ordered Regression Models of Educational Attainment

	Logit			Probit		Complementary Log Log		
(Educational Attainment: 1 = Less than High School, 2 = High School, 3 = Associate, 4 = College, 5 = Postgraduate)								
	Coef.	SE	OR	Coef.	SE	Coef.	SE	HR
Variables								
Age								
1 vs. 2–5	−0.007	0.007	1.126	−0.003	0.004	−0.006	0.006	1.106
2 vs. 3–5	0.013**	0.004	0.816	0.008**	0.003	0.008**	0.003	0.877
3 vs. 4–5	0.028**	0.009	0.640	0.016***	0.005	0.024**	0.008	0.677
4 vs. 5	0.013#	0.007	0.812	0.008*	0.004	0.010*	0.005	0.853
Female								
1 vs. 2–5	0.396#	0.220	0.673	0.191#	0.113	0.341#	0.184	0.711
2 vs. 3–5	0.148	0.126	0.863	0.083	0.077	0.082	0.088	0.921
3 vs. 4–5	−0.315	0.241	1.370	−0.160	0.132	−0.302	0.216	1.353
4 vs. 5	0.051	0.193	0.950	0.035	0.118	0.050	0.121	0.951
Race/ethnicity[a]								
Black								
1 vs. 2–5	−0.511	0.337	1.667	−0.277	0.175	−0.412	0.292	1.509
2 vs. 3–5	0.031	0.219	0.969	0.031	0.133	0.094	0.149	0.910
3 vs. 4–5	−0.555	0.363	1.742	−0.325	0.213	−0.451	0.306	1.570
4 vs. 5	−0.272	0.394	1.313	−0.172	0.243	−0.190	0.243	1.209
Latino								
1 vs. 2–5	−0.677*	0.327	1.968	−0.374*	0.177	−0.483#	0.275	1.621
2 vs. 3–5	−0.018	0.265	1.018	0.016	0.159	0.113	0.174	0.893
3 vs. 4–5	0.278	0.503	0.757	0.116	0.277	0.278	0.430	0.757
4 vs. 5	−1.199#	0.648	3.318	−0.692*	0.350	−0.608*	0.285	1.837
Other race/ethnicity								
1 vs. 2–5	−0.054	0.651	1.056	−0.047	0.328	−0.033	0.595	1.034
2 vs. 3–5	1.131**	0.376	0.323	0.669**	0.218	0.878**	0.309	0.416
3 vs. 4–5	1.166	0.748	0.312	0.614#	0.366	1.074	0.718	0.341
4 vs. 5	0.519	0.380	0.595	0.324	0.236	0.363	0.266	0.696
Parents' education								
Mother's education								
1 vs. 2–5	0.213***	0.039	0.448	0.118***	0.021	0.176***	0.033	0.516
2 vs. 3–5	0.088***	0.027	0.717	0.051**	0.016	0.046**	0.018	0.842
3 vs. 4–5	0.121*	0.048	0.635	0.066*	0.026	0.102*	0.042	0.680
4 vs. 5	0.055	0.041	0.813	0.034	0.025	0.037	0.025	0.871
Father's education								
1 vs. 2–5	0.113***	0.034	0.620	0.061***	0.018	0.095**	0.029	0.670
2 vs. 3–5	0.141***	0.022	0.549	0.085***	0.013	0.089***	0.015	0.685
3 vs. 4–5	0.064	0.040	0.762	0.042#	0.023	0.046	0.036	0.823
4 vs. 5	−0.063#	0.033	1.304	−0.038#	0.020	−0.038#	0.020	1.175

(Continued)

TABLE 4.3 *(Continued)*

Results from Nonparallel Continuation Ratio Ordered Regression Models of Educational Attainment

	(Educational Attainment: 1 = Less than High School, 2 = High School, 3 = Associate, 4 = College, 5 = Postgraduate)							
	Logit			Probit		Complementary Log Log		
	Coef.	SE	OR	Coef.	SE	Coef.	SE	HR
Cutpoints								
τ_1	0.584	0.648		0.348	0.339	0.028	0.544	
τ_2	3.498	0.450		2.072	0.265	1.708	0.285	
τ_3	1.855	0.818		1.066	0.459	1.134	0.685	
τ_4	1.099	0.706		0.711	0.433	0.489	0.444	
Model fit								
R^2_M		0.114			0.115		0.108	
R^2_O		0.147			0.144		0.126	
R^2_N		0.087			0.082		0.075	

Note: The results from significance tests are not reported for the cutpoints. SE = standard error, OR = odds ratio, HR = hazard ratio. X-standardized odds and hazard ratios are reported for continuous variables (age and parent's education), and un-standardized odds and hazard ratios are reported for binary variables (female and race/ethnicity). $N = 1286$.

[a] Reference = White.

[#]$p < 0.10$, [*]$p < 0.05$, [**]$p < 0.01$, [***]$p < 0.001$.

Based on the x-standardized odds and hazard ratios, father's education has the strongest effect among this group of variables.[*] A standard deviation increase in father's years of schooling is associated with a decrease in the hazard rate for a high school degree by a factor of 0.69 or 31%. Variables that have an overall positive association with educational attainment will tend to reduce the odds and hazard rate in the first few educational stages because they increase the chances that one will reach a higher level of educational attainment, such as a college or postgraduate degree.

In the third stage, the outcome is a binary indicator for an associate's degree versus a bachelor's degree or more, and the sample is limited to respondents with at least an associate's degree.[†] There are fewer significant coefficients in the associate stage. However, age and mother's education are significantly associated with decreases in the odds and hazard rate. The effects of these two variables are also nearly identical.

In the fourth and final stage, the outcome is a binary indicator for a college or bachelor's degree versus postgraduate degree based on a sample of respondents with a college degree or more. Only 494 of the original 1,286 respondents (38%) made the earlier transitions to this final stage, which may help explain the nonsignificant coefficients for most variables. However, this may also be due to a selection effect producing a more

[*] Odds and hazard ratios are already standardized in a sense in ordinal regression models due to the necessary scaling of coefficients based on residual variance. However, x-standardized odds ratios (see Long 1997) are commonly used to compare the relative magnitudes of these coefficients between continuous variables.

[†] By using a sequential model, we assume that respondents must pass through the associate stage in order to attain a bachelor's or postgraduate degree when in fact an associate's degree is not a prerequisite for a bachelor's degree. Therefore, researchers may choose to either include respondents with "some college" in the associate's degree category or combine the high school and associate's degree categories.

homogeneous sample in the final stage. Regardless, there are no variables in the college stage that have significant coefficients at the 0.05 alpha level using all three link functions. Age and race/ethnicity (Latino/White differences) are significant in two of the three models. Age is associated with a decrease in the hazard rate for attaining a college degree, whereas Latino identification is associated with an increase in the hazard rate. In addition, the coefficient for father's education is now marginally significant but in the opposite direction than in previous stages. A standard deviation increase in father's years of schooling is now associated with an increase in the hazard rate of attaining a college degree by a factor of 1.18 or 18%. Finally, the model fit statistics indicate that this group of variables reduces the deviance by 11%, explains 13%–15% of the ordinal variation, and explains 8%–9% of the nominal variation. Therefore, the nonparallel continuation ratio models are able to model ordinal relationships even though we relaxed the parallel regression restriction.

We examine the substantive effect sizes in more detail using average marginal effects, which we present in Table 4.4. These results are based on the nonparallel continuation ratio logit model with average discrete changes in parentheses. The average marginal effects are conditional in that they are specific to a particular equation or stage. Age has a relatively consistent average marginal effect across the educational stages after the initial stage. Although there is no effect of age on the probability of attaining less than a high school degree, it is associated with decreases in the probabilities in the remaining stages.

Parent's education also has relatively consistent effects across the educational stages. In the first three stages, increases in mother's and father's education are associated with decreases in the predicted probabilities of each educational level. For example, a standard deviation increase in mother's education is associated with an average decrease of 0.042 in the predicted probability of attaining less than a high school degree. The effect of mother's education is even stronger in the next two stages. However, father's education no longer has a significant effect in the third stage, and in the final stage the direction of the relationship changes. The average marginal effect of father's education on the predicted probability of a college degree is positive but only significant at the 0.10 level.

TABLE 4.4

Conditional Average Marginal Effects from a Nonparallel Continuation Ratio Logit Model of Educational Attainment

	Less than High School	High School	Associate	College
Age	0.000	−0.003**	−0.004**	−0.003*
	(0.008)	(−0.044)	(−0.058)	(−0.047)
Female	−0.026#	−0.032	0.046	−0.011
Race/ethnicity[a]				
Black	0.038	−0.007	0.090	0.058
Latino	0.053#	0.004	−0.038	0.221*
Other race/ethnicity	0.004	−0.233***	−0.125*	−0.119
Parents' education				
Mother	−0.014***	−0.019***	−0.018*	−0.012
	(−0.042)	(−0.072)	(−0.059)	(−0.046)
Father	−0.007***	−0.031***	−0.009	0.014#
	(−0.028)	(−0.128)	(−0.037)	(0.056)

Note: The numbers in parentheses are average standard deviation discrete changes.
[a] Reference = White.
#$p < 0.10$, *$p < 0.05$, **$p < 0.01$, ***$p < 0.001$.

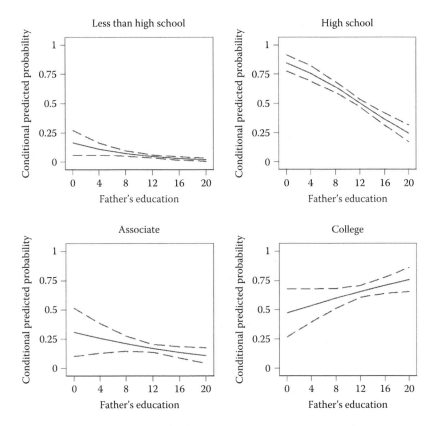

FIGURE 4.2
The conditional effect of father's education on educational attainment from a nonparallel continuation ratio logit model.

We examine the effects of father's education on the conditional predicted probabilities in more detail in Figure 4.2. The graphs plot the conditional predicted probabilities from the four educational stages across the range of father's education. Father's education is most strongly associated with changes in the conditional probabilities of a high school or college degree. Among the sample of respondents with a college or postgraduate degree, the predicted probability of a college degree increases by 0.28 across the range of father's years of schooling. However, the confidence interval is much wider for college than high school, which is due in part to the smaller sample size in the final educational stage. The effect of father's education on the conditional predicted probability of a high school degree is approximately twice as strong. The predicted probability decreases by 0.60 across the range of father's education. Overall, the results in this figure indicate that father's education increases the probability that one will make the first few educational transitions, but it decreases the probability that one will make the final transition from a college degree to a postgraduate degree.

The results from this example highlight the usefulness of the nonparallel continuation ratio model for analyzing ordinal outcomes that are the result of a sequential process. Educational attainment is a sequence of stages and the nonparallel model allows the effects of each independent variable to freely vary across the stages. The coefficients for most of the variables in the model vary considerably across the educational stages, which means the parallel regression assumption is too restrictive. The sample size in this example is large ($N = 1{,}286$). Therefore, the added model complexity with 21 additional slopes should not be a problem. Additionally, the nonparallel continuation ratio model does not suffer from the potential problem of negative predicted probabilities that is inherent in the partial and nonparallel cumulative models. However, there are other concerns in sequential models due to the conditional nature of the relationships. Unfortunately, we cannot completely distinguish changes in "true effects" from changes in unobserved heterogeneity across the transitions (Mare 2006). Researchers need to keep this in mind when interpreting the stage-specific results and making claims regarding the theoretical implications of changes in coefficients in later stages.

4.3 The Nonparallel Adjacent Category Model

The nonparallel adjacent category model is well suited for ordinal outcomes consisting of discrete categories that are of substantive interest to the researcher. It is the only class of ordered regression models that focuses on comparisons of two outcome categories in the binary cutpoint equations. It is also the only model we consider that is appropriate for both ordinal and nominal outcomes. It does not impose the parallel regression assumption, and reordering the outcome categories does not change the model in a meaningful way. The nonparallel adjacent category model has several advantages over the nonparallel cumulative and continuation ratio models. First, the model does not have the internal inconsistency and potential for negative predicted probabilities that is found in the nonparallel cumulative model. Second, relaxing the parallel regression assumption does not require the researcher to discard the latent variable motivation, which is another limitation of the nonparallel cumulative model. The nonparallel adjacent category logit model is equivalent to the multinomial logit model, which has a latent variable motivation within the random utility model framework (McFadden 1973). Finally, one does not have to assume that the outcome is the result of an irreversible sequential decision-making process, which makes the continuation ratio model inappropriate for many ordinal outcomes.

However, the nonparallel adjacent category model requires the independence of irrelevant alternatives (IIA) assumption, which is difficult to test. The IIA assumption requires that the ratio of the probabilities for two categories (e.g., 1 vs. 2 or 2 vs. 3) are not affected by the addition of a new category (Long 1997, pp. 182–183). Unfortunately, the size properties for formal tests of the IIA assumption are dependent upon the structure of the data and therefore not useful for applied work (see Cheng and Long 2007). In the absence of formal IIA tests, researchers interested in using the nonparallel adjacent category model should be confident that the outcome categories are "distinct" and "weighed independently" by respondents (McFadden 1973). The categories should be dissimilar enough that they cannot be reasonably considered as substitutes for one another (Amemiya 1981, p. 1517; Cheng and Long 2007, p. 598). For more details on the IIA assumption, see the discussion in Chapter 2.

The nonparallel adjacent category model is a nonlinear adjacent probability model that relaxes the parallel regression assumption for every variable. The general form of the equation for the nonlinear adjacent probability model is

$$\Pr(y = m | y = m \text{ or } y = m + 1, \mathbf{x}) = F(\tau_m - \mathbf{x}\boldsymbol{\beta}_m) \qquad (1 \leq m < M) \qquad (4.4)$$

where F represents the CDF for the logit or probit model, τ_m are cutpoints, $\mathbf{x}$ is a vector of independent variables, and $\boldsymbol{\beta}_m$ is a vector of coefficients that freely vary across cutpoint equations.

The original ordinal outcome is recoded into a series of $M - 1$ binary outcomes corresponding to each cutpoint equation. If $M = 4$, then the nonparallel adjacent category model estimates the following three binary cutpoint equations simultaneously:

Cutpoint equation #1: $\Pr(y = 1 | y = 1 \text{ or } y = 2, \mathbf{x}) = F(\tau_1 - \mathbf{x}\boldsymbol{\beta}_1)$
Cutpoint equation #2: $\Pr(y = 2 | y = 2 \text{ or } y = 3, \mathbf{x}) = F(\tau_2 - \mathbf{x}\boldsymbol{\beta}_2)$
Cutpoint equation #3: $\Pr(y = 3 | y = 3 \text{ or } y = 4, \mathbf{x}) = F(\tau_3 - \mathbf{x}\boldsymbol{\beta}_3)$

The sample size in each cutpoint equation changes depending on the marginal distribution for the dependent variable. If the cases are evenly distributed across the categories, then the sample sizes will be the same across the cutpoint equations. However, if a majority of the cases are clustered in one or two categories, then the cutpoint equation sample sizes will vary substantially. The values for the cutpoints (τ_m) and the coefficients for the independent variables ($\boldsymbol{\beta}_m$) vary across equations, whereas the values for the independent variables ($\mathbf{x}$) remain constant across equations.

The cutpoint equations are based on comparisons of adjacent categories, but it is also possible to set up the model as a baseline model equivalent to multinomial logit. Therefore, the nonparallel adjacent category logit model is also closely related to the stereotype logit model. A stereotype logit model with the maximum number of dimensions is equivalent to the nonparallel adjacent category logit model.

4.3.1 Example: Nonparallel Adjacent Category Models of Welfare Spending Attitudes

We will illustrate the nonparallel adjacent category model using the welfare spending attitudes example from the 2012 GSS that we introduced in Chapter 1. We present coefficients and odds ratios from the nonparallel adjacent category logit and probit models in Table 4.5.[*] The odds ratios are x-standardized for the continuous variables (age, income, education, ideology, and racial attitude). There are three outcome categories (too much, about right, and too little) and therefore two cutpoint equations. The parallel regression assumption is relaxed for the entire model, which means that there are two coefficients for every independent variable.

In the first equation, the sample is limited to respondents who either oppose welfare spending ("too much") or express mixed support ("about right"), and in the second equation, the sample is limited to mixed support and support ("too little"). The nonparallel model allows each variable to have different effects on opposition versus mixed support and mixed support versus support for welfare spending. The factors associated with support for welfare spending are not necessarily also related to opposition (Fullerton and Dixon 2009). The coefficients and significance levels vary across equations for nearly every variable in the model, which suggests that the nonparallel model may be particularly useful for this example.

The most consistent pattern in the results is a stronger association in the first cutpoint equation (opposition vs. mixed support) than in the second cutpoint equation

---

[*] The probit model requires the IIA assumption as well. We do not consider the alternative-specific multinomial probit model, which allows the errors to be correlated across equations. Additionally, the complementary log log link function is not used for adjacent or baseline category models.

TABLE 4.5

Results from Nonparallel Adjacent Category Ordered Regression Models of Welfare Spending Attitudes

	(Welfare Spending Attitude: 1 = Too Much, 2 = About Right, 3 = Too Little)				
	Logit			Probit	
	Coef.	SE	OR	Coef.	SE
Variables					
Age					
1 vs. 2	0.017*	0.007	0.752	0.013*	0.006
2 vs. 3	−0.005	0.009	1.091	−0.004	0.007
Female					
1 vs. 2	−0.329	0.245	1.390	−0.250	0.198
2 vs. 3	0.430	0.319	0.650	0.328	0.235
Income					
1 vs. 2	−0.295*	0.118	1.400	−0.243**	0.094
2 vs. 3	−0.169	0.139	1.212	−0.118	0.104
Education					
1 vs. 2	0.083#	0.049	0.795	0.067#	0.040
2 vs. 3	−0.141*	0.062	1.481	−0.114*	0.046
Ideology					
1 vs. 2	−0.005	0.098	1.008	−0.005	0.079
2 vs. 3	−0.228#	0.126	1.391	−0.172#	0.092
Party identification[a]					
Democrat					
1 vs. 2	1.266***	0.367	0.282	1.051***	0.295
2 vs. 3	0.396	0.565	0.673	0.156	0.383
Independent					
1 vs. 2	1.124***	0.329	0.325	0.927***	0.264
2 vs. 3	0.432	0.539	0.649	0.203	0.356
Racial attitude					
1 vs. 2	−0.266*	0.128	1.309	−0.215*	0.104
2 vs. 3	−0.202	0.159	1.227	−0.140	0.119
Cutpoints					
τ_1	−1.132	1.459		−0.937	1.166
τ_2	−4.211	1.769		−3.275	1.346
Model Fit					
R_M^2		0.099		0.100	
R_O^2		0.131		0.131	
R_N^2		0.105		0.105	

Note: The results from significance tests are not reported for the cutpoints. SE = standard error, OR = odds ratio. X = standardized odds ratios are reported for continuous variables (age, income, education, ideology, and racial attitude), and un-standardized odds ratios are reported for binary variables (female and party identification). $N = 392$.

[a] Reference = Republican.

#$p < 0.10$, *$p < 0.05$, **$p < 0.01$, ***$p < 0.001$.

(mixed support vs. support). For example, a standard deviation increase in age is associated with a decrease in the odds of welfare opposition by a factor of 0.75 or 25%. The coefficient for age is in the opposite direction but nonsignificant in the second equation. All else equal, older respondents have lower odds of opposing welfare spending than younger respondents, but this does not translate into higher levels of support. We also see coefficients with opposite signs for sex, but the sex differences are nonsignificant in both equations.

Party identification and racial attitude also have larger coefficients in the equation for welfare opposition than in the equation for mixed support. Identifying as a Democrat (vs. as a Republican) is associated with decreases in the odds of welfare opposition by a factor of 0.28 or 72%. The odds ratios for party identification are closer to one and nonsignificant in the second equation. Negative racial attitude is positively associated with opposition and mixed support, but it is also only significant in the first equation.

Education and ideology, however, display stronger associations with mixed support than opposition to welfare spending. Conservative ideology is positively associated with the odds of mixed support, but it has little to no association with opposition. The difference is even more pronounced for education. A standard deviation increase in education is associated with a decrease in the odds of opposition by a factor of 0.80 but with an increase in the odds of mixed support by a factor of 1.48.

The nonparallel adjacent category model allows us to uncover this asymmetric relationship between education and welfare attitudes. Imposing the parallel regression assumption for education would result in a nonsignificant negative coefficient by averaging the marginally significant positive coefficient in the first equation with the significant negative coefficient in the second equation. This would lead us to conclude that there are no net educational differences in welfare attitudes. The results from the nonparallel model challenge this conclusion and instead support the idea that highly educated respondents are more likely to express mixed support for welfare opposition than the more extreme positions of opposition or support. According to the model fit statistics, the variables reduce the model deviance by 10%, explain 13% of the ordinal variation, and explain 11% of the nominal variation. The variables explain nearly as much nominal as ordinal variation, which is what we should expect given that the nonparallel adjacent category model is appropriate for both nominal and ordinal outcomes.

We examine the substantive effect sizes for each variable using average marginal effects based on the nonparallel adjacent category logit model (see Table 4.6). Party identification has the most consistent effects on the predicted probabilities across outcome categories. On average, identifying as a Democrat (vs. as a Republican) is associated with a 0.29 decrease in the predicted probability of opposition, a 0.15 increase in the predicted probability of mixed support, and a 0.14 increase in the predicted probability of support for welfare spending. Income and racial attitude also have relatively strong effects on the probabilities for welfare opposition and support. Education is the only significant predictor that has its strongest association with mixed support.

We plot the predicted probabilities across the range of negative racial attitude in Figure 4.3 in order to examine its association with welfare attitudes in more detail. The racial attitude measure is based on the respondent's evaluation of Blacks' work ethic (1 = hard working and 7 = lazy). As one's racial attitude increases from extremely positive to extremely negative, the predicted probability of opposition to welfare spending increases by 0.46, which is a very substantial increase. Racial attitude also has notable but weaker effects on the predicted probabilities for support and mixed support. As the respondent's racial attitude becomes increasingly negative, the predicted probability of support and mixed support decreases. Respondents with strong, positive racial attitudes

TABLE 4.6

Average Marginal Effects from a Nonparallel Adjacent
Category Logit Model of Welfare Spending Attitudes

	Too Much	About Right	Too Little
Age	−0.003*	0.003*	0.000
	(−0.055)	(0.048)	(0.007)
Female	0.042	−0.076	0.034
Income	0.074***	−0.032	−0.042**
	(0.084)	(−0.040)	(−0.044)
Education	−0.008	0.021*	−0.013#
	(−0.026)	(0.060)	(−0.034)
Ideology	0.016	0.014	−0.030*
	(0.022)	(0.018)	(−0.040)
Party identification[a]			
Democrat	−0.289***	0.151#	0.139#
Independent	−0.257***	0.130#	0.128*
Racial attitude	0.070**	−0.026	−0.044*
	(0.070)	(−0.029)	(−0.041)

Note: The numbers in parentheses are average standard deviation
discrete changes.
[a] Reference = Republican.
#$p < 0.10$, *$p < 0.05$, **$p < 0.01$, ***$p < 0.001$.

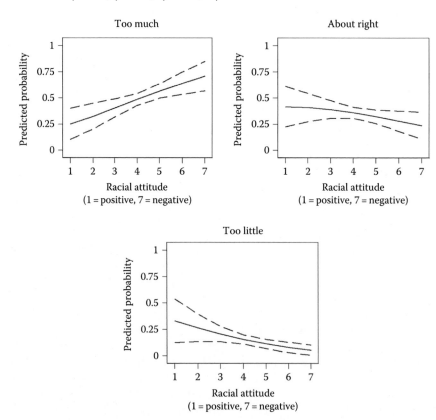

FIGURE 4.3
The effect of racial attitudes on welfare spending attitudes from a nonparallel adjacent category logit model.

are predicted to express either support or mixed support for welfare spending ($p = 0.75$), whereas respondents with strong, negative racial attitudes are predicted to oppose welfare spending ($p = 0.71$). Thus, racial attitudes are closely linked to welfare attitudes, and this supports previous research on the "racialization of welfare" in the United States (Gilens 1999; Fullerton and Dixon 2009).

On balance, the results from this example highlight the flexible nature of the nonparallel adjacent category model. This is a method for nominal outcomes that is also appropriate for ordinal outcomes if the parallel regression assumption is too restrictive. The main limitation associated with this method is the IIA assumption, which requires that the ratio of the probabilities for two categories is unaffected by the introduction of additional alternatives. This is a necessary assumption given the conditional nature of the adjacent probabilities. The samples for each cutpoint equation are limited to two adjacent categories. If the categories are too similar, then it is reasonable to assume that respondents view them as substitutes; therefore, IIA is not a reasonable assumption. In that case, other methods such as alternative-specific multinomial probit or mixed logit may be used (see Train 2009). The IIA assumption is reasonable for this example because the three outcomes represent three distinct views of welfare spending.

Relaxing the parallel regression assumption for the entire model allows us to uncover several asymmetrical relationships. For example, age is negatively associated with the probability of opposition to welfare spending, but there is no association with spending support. Additionally, education is negatively associated with the probability of both opposition and support. All else equal, highly educated respondents are more likely to take the middle position of mixed support. The application of the parallel regression assumption would lead one to conclude that education is statistically unrelated to welfare attitudes because it averages the negative and positive coefficients from the cutpoint equations. We also uncovered this more nuanced relationship in the previous chapter using the partial adjacent category model. The advantage of the nonparallel model is that it simplifies the process of relaxing the parallel regression assumption. If the assumption is violated for at least one variable, then it is relaxed for the entire model. However, the nonparallel model is also the least parsimonious ordered regression model.

4.4 Practical Issues in the Estimation of Nonparallel Models

Estimating nonparallel models is a relatively simple and straightforward process. The nonparallel adjacent category logit model is simply a reparameterization of the more familiar multinomial logit model, which is available in standard statistical software packages. Although software options for the nonparallel continuation ratio model are more limited, due to the assumption of independent errors across equations one may estimate this model as a series of separate binary regression models (Fienberg 1980). The sample sizes are progressively smaller in later stages because those equations are only based on respondents that made all of the earlier transitions. However, there are user-written commands that estimate the cutpoint equations simultaneously (e.g., Buis 2007; Yee 2010).[*]

---

[*] We also developed our own Stata macros for the 12 ordered regression models in this book, which are available with examples online: http://sociology.okstate.edu/faculty-staff-directory/faculty-directory/dr-andrew-fullerton.

It is also fairly easy to estimate the nonparallel cumulative model, but one must either rely on user-written commands, such as gologit2 in Stata (Williams 2006) or VGAM in R (Yee 2010), or on the "person/threshold approach" (Cole et al. 2004). The latter approach requires data expansion from the person-level to the person/threshold-level by creating replicate records for each additional cutpoint equation (see Chapter 2). For the cumulative model, each respondent contributes $M - 1$ observations in the person/threshold file, where M represents the number of outcome categories. The binary outcome and indicators for each "threshold" or cutpoint equation vary across the replicate datasets, whereas the values for the independent variables remain fixed. One relaxes the parallel regression assumption by including interaction terms in the person/threshold file between the binary cutpoint variables and the independent variables. After the person/threshold file is constructed, one estimates the model as a binary regression model. The presence of multiple observations per person creates dependence among the errors, which requires the use of methods such as generalized estimating equations or a binary regression model with robust standard errors (see Fullerton and Xu 2012).

4.5 Conclusions

Nonparallel models are the least restrictive ordered regression models that we consider in this book. They do not require the parallel regression assumption, which is often violated in practice. If the coefficients for a given variable are relatively constant across the cutpoint equations, then the nonparallel model will simply be less efficient because it requires additional parameters to produce approximately the same predicted probabilities. If there are at least 300–400 cases and 10–20 cases per parameter, then the added model complexity is not a serious concern. However, researchers need to keep in mind the limitations associated with the three classes of nonparallel models, including negative predicted probabilities in the cumulative model, the potential for sample selection bias and scaling effects in the continuation ratio model, and the IIA assumption in the adjacent model. Although they are based on comparisons of different sets of categories in the cutpoint equations, the three nonparallel models may produce very similar predicted probabilities. This is the case for the empirical examples in this chapter.

In the empirical examples, we saw that relaxing the parallel regression assumption for every independent variable led to very different substantive conclusions. In the cumulative models of self-rated health, the nonparallel models revealed several asymmetrical relationships that are obscured by the parallel regression assumption. For example, marital status is negatively associated with the probability of poor health, but it has little to no association with good or excellent health. We found a similar pattern for the effects of education in the adjacent category models of welfare attitudes. In the parallel models, education is statistically unrelated to welfare attitudes, but in the nonparallel model, it is significant and positively associated with the probability of mixed support for welfare spending. In other cases, we found that the effects for some variables were significant in the parallel and nonparallel models, but nonparallel models revealed that the effects grew weaker or stronger across equations. Good examples of this are the weakening effects of parents' education in later stages of educational attainment.

4.5.1 Guidelines for Choosing a Parallel, Partial, or Nonparallel Ordered Regression Model

In the previous chapter, we suggested that researchers should consider two main factors when choosing a parallel or partial ordered regression model. Parallel models are preferred if there is no ambiguity in the ordering of outcome categories, and the parallel regression assumption is a reasonable assumption for each independent variable. This applies when we consider nonparallel models as well.

If the parallel regression assumption is too restrictive for one or more variables in the model, then the researcher must decide whether to relax the assumption for the entire model or on a variable-specific basis. Nonparallel models are best suited for ordinal outcomes with some ambiguity in the outcome category order. For instance, measures such as subjective social class (upper, middle, working, and lower) are often treated as ordinal, but this requires the potentially problematic assumption that "working" class is located between the "lower" and "middle" classes. Additionally, midpoint categories in Likert scales, such as "neither agree nor disagree," introduce a degree of ambiguity given that the responses in these categories may also be a form of nonresponse (Sturgis et al. 2014). Respondents may choose this category if they do not have an opinion on the topic or do not wish to share it with the interviewer. The nonparallel adjacent category logit model is arguably the best model to use if there is some question as to whether the outcome is ordinal or nominal because it is equivalent to the multinomial logit model for nominal outcomes. Unlike the nonparallel cumulative and continuation ratio models, reordering the outcome categories does not affect the results in the nonparallel adjacent category model. However, this also indicates that the nonparallel adjacent category model is the least efficient model for ordinal outcomes that we consider in this book because it requires the most parameters, and it does not take the category order into account.

Partial and nonparallel ordered regression models are likely to produce very similar results. Both models relax the parallel regression assumption for a subset of variables that have coefficients that vary across equations. The nonparallel model is less efficient than the partial model because it also relaxes the parallel assumption for the remaining variables in the model, which may have coefficients with little to no variation across equations. The partial and nonparallel models are likely to have very similar predicted probabilities, but the nonparallel model is less efficient because it requires more parameters. For example, if the parallel assumption is reasonable for six of the eight independent variables and there are three cutpoint equations, then the partial model will have 12 fewer parameters than the nonparallel model (12 vs. 24 slopes).

As a result, one should consider using a partial model rather than a nonparallel model if parsimony is an important consideration. Ideally, one should have at least 10–20 cases per parameter. Therefore, ordered regression models in smaller samples need to be more parsimonious. One may obviously increase the level of parsimony by limiting the number of independent variables, but the application of the parallel regression assumption is equally important. In large samples, parsimony is less important, and the decision regarding whether to use a partial or nonparallel model may come down to practical issues such as software availability. Nonparallel models are easier to estimate using standard statistical software packages. Partial models may require the researcher to impose the constraints manually in the estimation process (for exceptions, see Williams 2006; Yee 2010; Pfarr et al. 2011).

The choice of a particular ordered model is closely related to formal and informal tests of the parallel regression assumption. One may compare the relative model fit of the nonparallel and parallel models as an omnibus test of the parallel regression assumption for the entire model. Alternatively, one may compare the relative model fit of partial and parallel models in order to test the parallel regression assumption for one variable or a group of variables. Thus, partial and nonparallel models are useful alternatives and important methods for testing the key assumption in parallel ordered regression models. In the next chapter, we introduce several formal and informal tests of the parallel regression assumption and discuss the use of these tests in the selection of an appropriate ordered regression model.

4.6 Appendix

4.6.1 Stata Codes for Nonparallel Ordered Logit Models

Nonparallel Cumulative Logit Model of Self-Rated Health

```
gologit2 ghealth age agesq female black hisp2 othrace educ employed lninc
    married
```

Nonparallel Continuation Ratio Logit Model of Educational Attainment

```
seqlogit degree5 age female black hisp2 othrace maeduc paeduc, tree(1 : 2
    3 4 5 , 2 : 3 4 5 , 3 : 4 5 , 4 : 5)
```

Nonparallel Adjacent Category Logit Model of Welfare Spending Attitudes

```
mlogit natfare2 age female lninc educ polviews democrat indep workblks,
    base(2)
```

4.6.2 R Codes for Nonparallel Ordered Logit Models

Nonparallel Cumulative Logit Model of Self-Rated Health

```
npLogit <- vglm(as.ordered(ghealth) ~ age + agesq + female
            + black + hisp2 + othrace
            + educ + employed + lninc + married,
            cumulative(link="logit",
                    parallel = FALSE,
                    reverse = TRUE),
            etastart = predict(poLogit),
            trace = TRUE, # Rank = 1,
              data = mydta)
summary(npLogit)
```

Nonparallel Continuation Ratio Logit Model of Educational Attainment

```
crNpLogit <- vglm(as.ordered(degree5) ~ age + female + black
              + hisp2 + othrace + maeduc + paeduc,
              cratio(link="logit",
                    parallel = FALSE,
                    reverse = FALSE),
```

```
            etastart = predict(crPoLogit),
            trace = TRUE, # Rank = 1,
            data = mydta)
summary(crNpLogit)
```

Nonparallel Adjacent Category Logit Model of Welfare Spending Attitudes

```
acNpLogit <- vglm(as.ordered(natfare2) ~ age + female + educ
            + lninc + polviews + democrat
            + indep + workblks,
            acat(link="loge",
              parallel = FALSE,
              reverse = FALSE),
            trace = TRUE, # Rank = 1,
            data = mydta)
summary(acNpLogit)
```

5

Testing the Parallel Regression Assumption

Parallel ordered regression models share a common assumption, which is that the effects of independent variables do not vary across cutpoint equations. We refer to this as the parallel regression assumption, and for logit models it is also known as the "proportional odds" assumption (McCullagh 1980). We provide an illustration of the parallel regression assumption for cumulative models in Figure 5.1. In the graph on the left, the cumulative probability curves are parallel, whereas in the graph on the right they are not. The parallel regression assumption is very useful because it produces a relatively parsimonious model and restricts the coefficients to ensure ordinality in the relationships. However, the assumption is often violated in practice.

In this chapter, we discuss the parallel regression assumption, present a variety of tests of this assumption, address the strengths and weaknesses of each test, and provide suggestions for applied researchers. In formal tests, a significant statistic indicates that the variation in coefficients across cutpoint equations is statistically significant. However, this does not necessarily mean that there is a substantial amount of variation in effect sizes across equations or that relaxing the assumption will affect the substantive conclusions. One general point that we will emphasize throughout the chapter is that a statistically significant test statistic does not necessarily imply that the parallel regression assumption is not reasonable for a particular variable or model. We recommend using several different tests in order to determine whether or not to apply the parallel regression assumption. At the end of each empirical example, we will summarize the results from all of the formal and informal tests and draw conclusions about the preferred ordered regression models. Most formal tests of the parallel regression assumption are based on a Wald or a likelihood ratio (LR) test. We will begin by briefly reviewing these tests.

5.1 Wald and LR Tests

The Wald test (Wald 1943) was designed for a composite set of parameters in large samples. In a special case where there is a single parameter in the hypothesis, for example, $H_0 : \hat{\beta} = \beta_0$ versus $H_1 : \hat{\beta} \neq \beta_0$, the Wald test statistic can be computed using the z statistic:

$$z = \left(\hat{\beta} - \beta_0\right) / \mathrm{SE}\left(\hat{\beta}\right) \tag{5.1}$$

Alternatively, one can square it and use a χ^2 test to get the same results. When a set of linear constraints is imposed on parameter estimates (e.g., H_0: $T\theta = c$), the Wald test statistic can be constructed as follows:

$$W = \left(T\hat{\theta} - c\right)^{\mathrm{T}} \left(T\hat{V}\left(\hat{\theta}\right)T^{\mathrm{T}}\right)^{-1} \left(T\hat{\theta} - c\right) \sim \chi_r^2 \tag{5.2}$$

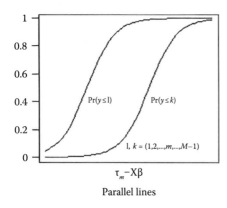 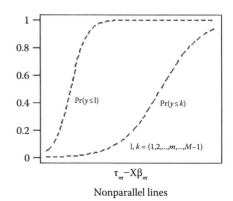

FIGURE 5.1
The parallel regression assumption.

where T is a linear transformation matrix of $\boldsymbol{\theta}$, $\mathbf{c}$ is a column of constants, r is the number of linearly independent constraints, $\hat{\boldsymbol{\theta}}$ is a vector of unconstrained maximum likelihood (ML) estimates of $\boldsymbol{\theta}$, and $\hat{V}\left(\hat{\boldsymbol{\theta}}\right)$ is the estimated variance–covariance matrix of $\hat{\boldsymbol{\theta}}$.

Unlike the Wald test, which only requires the estimation of the unconstrained model, the LR test (Wilks 1938) requires the estimation of both constrained and unconstrained models. For tests of the parallel regression assumption, the unconstrained model is typically the nonparallel ordered regression model with slopes that vary freely across cutpoint equations for every variable, or what is commonly known as the generalized ordered regression model (Williams 2006). In the constrained model, the effects of one or more independent variables are restricted to remain constant across cutpoint equations. For example, if there are three cutpoint equations, then the following two constraints are placed on the slopes in the constrained model: $\beta_{1k} = \beta_{2k}$ and $\beta_{1k} = \beta_{3k}$. The constrained model is "nested" within the unconstrained model, which is a requirement for the LR test. The LR test is a function of the difference in the log likelihoods for the constrained and unconstrained models:

$$LR = 2(LL_U - LL_C) \tag{5.3}$$

where LL_U and LL_C are the log likelihoods for the unconstrained and constrained models, respectively. The null hypothesis implied by the previous example is H_0: $\beta_{1k} = \beta_{2k} = \beta_{3k}$. Assuming the null hypothesis is true, the LR test statistic is asymptotically distributed as a chi-square statistic with the degrees of freedom (df) equal to the number of constraints. In this example, $df = 2$. The Wald and LR tests are asymptotically equivalent, but there are differences between the test statistics in small, finite samples. The LR test statistic is simple to calculate, but it requires the estimation of the unconstrained and constrained models. Historically, this posed problems when using the LR test to assess the parallel regression assumption in cumulative models because standard software was unavailable for the estimation of nonparallel cumulative models. However, it is now possible to estimate these models using several different programs (e.g., see Williams 2006; Yee 2010). One of the first tests of the parallel regression assumption, the score test, only requires the estimation of the constrained model. We consider this test in the next section.

5.2 The Score Test

Rao (1948) proposed the score test as an alternative to the Wald and LR tests. The test is called the "score" test because the test statistic involves the score function, which is the gradient or first partial derivative of the log-likelihood function of the unconstrained model evaluated with the constrained estimates. Rao's original work, however, remained in obscurity for a few decades, during which time it was reinvented by Silvey (1959) as a "Lagrange multiplier" (LM) test. Breusch and Pagan (1980) then rediscovered the score test in the context of the general econometric model specification search.

Unlike the Wald and LR tests, the score test only estimates the constrained model and assesses whether the model performs well enough in maximizing the log-likelihood function based on the score function. In a simple one-parameter case with H_0: $\theta = \theta_r$ using maximum likelihood estimation (MLE), the score test statistic can be constructed as

$$\frac{\left(g\left(\hat{\theta}_r\right)\right)^2}{\mathrm{Var}\left(g\left(\hat{\theta}_r\right)\right)} \sim \chi^2(1) \tag{5.4}$$

where $\hat{\theta}_r$ is a parameter estimate of θ under the null and $g\left(\hat{\theta}_r\right)$ is the score function (for more details, see Buse 1982). When there is a composite set of constraints, H_0: $\boldsymbol{\theta} = \boldsymbol{\theta}_r$, the scalar version of the score test statistic can be easily generalized to a multivariate matrix form as follows:

$$g\left(\hat{\theta}_r\right)^{\mathrm{T}} I\left(\hat{\theta}_r\right)^{-1} g\left(\hat{\theta}_r\right) \sim \chi_r^2 \tag{5.5}$$

with r restrictions. The score test is typically set up as an "omnibus" test in which the regression slopes are constrained to be constant across cutpoint equations for every independent variable. The χ^2 for this omnibus test has $K(M - 2)$ degrees of freedom, where K is the number of independent variables and M is the number of outcome categories. A statistically significant score test statistic indicates that the parallel regression assumption is violated for the entire model (for more details, see the Appendix).

5.2.1 Example: Cumulative Logit Model of General Self-Rated Health

To provide an empirical illustration, we will present the results from a score test of the parallel regression assumption for the parallel cumulative logit model of general self-rated health from Chapter 2. To simplify the presentation of results, we will focus on the logit link function for the examples in this chapter. There are four categories for health (poor, fair, good, and excellent) and 10 independent variables in each model. Hence, $M = 4$ and $K = 10$. Therefore, the degrees of freedom for the score test is $df = 10 \times (4 - 2) = 20$. The null hypothesis for an omnibus test of the parallel regression assumption is H_0: $\boldsymbol{\beta}_1 = \boldsymbol{\beta}_2 = \boldsymbol{\beta}_3$. The score test for this example is statistically significant ($\chi^2 = 37.85$, $df = 20$, $p = 0.009$), which indicates that it is not reasonable to assume that all of the regression slopes are constant across cutpoint equations.[*] However, this result may be due to variation in the slopes of one or two variables rather than in all 10 variables across the cutpoint equations.

---

[*] The score test is available in Stata using the "oparallel" command (Buis 2013). It is also reported in the default output for the parallel cumulative logit model in some programs, such as SAS. Currently, the score test is not available for the continuation ratio or adjacent category models.

5.3 The Brant Test

Brant (1990) proposed a weak version of the Wald test to assess the parallel regression assumption. The logic of the Brant test is very much like that of the Wald test of the parallel regression assumption. To set up the Brant test, one needs to first estimate a set of $M - 1$ separate binary regression models for an ordered response variable $y = 1,...,m,...,M$; each model is a binary logistic regression of $w_m (m = 1, 2,...,M - 1)$, which is set to be one if $y > m$ and zero if $y \leq m$, on $\mathbf{X}$ with K explanatory variables that are the same as those in the parallel model (Brant 1990, p. 1172; Long 1997, p. 144).[*] The Brant test compares regression slopes across cutpoint equations either collectively or individually for the independent variables.

Under the null hypothesis that $\boldsymbol{\beta}_1 = ... =\boldsymbol{\beta}_m ... \boldsymbol{\beta}_{M-1}$, an omnibus test of the parallel regression assumption using the Wald statistic can be constructed as

$$\left(\mathbf{D}\hat{\boldsymbol{\beta}}_w\right)^{\mathrm{T}} \left(\mathbf{D}\mathbf{V}\left(\hat{\boldsymbol{\beta}}_w\right)\mathbf{D}^{\mathrm{T}}\right)^{-1} \left(\mathbf{D}\hat{\boldsymbol{\beta}}_w\right) \sim \chi^2_{(M-2)K} \tag{5.6}$$

where $\hat{\boldsymbol{\beta}}_w$ denotes the ML estimates from the full set of $M - 1$ binary regression of w_m on $\mathbf{X}$, excluding the intercept, and $\hat{\boldsymbol{\beta}}_w = \left(\hat{\boldsymbol{\beta}}^{\mathrm{T}}_{w_1},....,\hat{\boldsymbol{\beta}}^{\mathrm{T}}_{w_m},....,\hat{\boldsymbol{\beta}}^{\mathrm{T}}_{w_{M-1}}\right)^{\mathrm{T}}$ (Brant 1990, p. 1173; Long 1997, p. 144). According to Brant (1990, p. 1172, 1177), $\hat{\boldsymbol{\beta}}_w$ follows an asymptotically multivariate-normal distribution, its expectation approximates the population parameter vector, $\boldsymbol{\beta}_0$, and its asymptotic variance–covariance, $\mathrm{V}\left(\hat{\boldsymbol{\beta}}_w\right)$; $\mathbf{D}$ is a $(M - 2)K \times (M - 1)K$ block design matrix that sets up the comparisons among slopes for the same variables across cutpoint equations (Brant 1990, p. 1173; Long 1997, p. 144). This design matrix can also be modified to test the assumption for one variable or for a subset of variables. The χ^2 for the omnibus test has $K(M - 2)$ degrees of freedom, and the variable-specific test has $M - 2$ degrees of freedom. A statistically significant Brant test result indicates that the parallel regression assumption is violated.

5.3.1 Example: Cumulative Logit Model of General Self-Rated Health

Using the "brant" command in Stata (Long and Freese 2014), we present the results from a Brant test of the parallel regression assumption for the parallel cumulative logit model of general self-rated health in Table 5.1.[†] The results in the top row correspond to an omnibus test of the parallel regression assumption. At the conventional alpha level of 0.05, we can reject the null hypothesis that all regression coefficients are equal across the cutpoint equations. The Brant tests for individual variables also show that there are 3 out of 10 variables for which the parallel regression assumption is violated at the 0.05 level, including age, age-squared, and employed. The Brant tests are nonsignificant for the remaining variables, which suggests that the parallel assumption is reasonable for those variables. In summary, the parallel regression assumption does not hold for this model because the effects of age, age-squared, and employed vary significantly across cutpoint equations.

---

[*] The following discussion pertains to the logit link only. For other links, see Brant (1990, p. 1178).

[†] Currently, the Brant test is only available for the cumulative logit model.

TABLE 5.1

Results from the Brant Test of the Parallel Regression Assumption in the Parallel Cumulative Logit Model of General Self-Rated Health

	$y > 1$	$y > 2$	$y > 3$	χ^2	*p*-value
Omnibus test ($df = 20$)				38.480	0.008
Age	−0.264	−0.085	−0.067	9.180	0.010
Age^2	0.002	0.001	0.000	8.450	0.015
Female	−0.015	−0.149	0.077	1.900	0.387
Race/ethnicity					
Black	−0.159	0.224	0.016	1.630	0.443
Latino	−0.088	−0.419	−0.398	0.520	0.773
Other race/ethnicity	0.382	−0.408	−0.647	0.920	0.630
Education	0.071	0.104	0.120	0.790	0.673
Employed	1.691	0.784	0.237	16.490	0.000
Log income	0.295	0.285	0.282	0.010	0.996
Married	0.753	0.257	0.076	3.520	0.172

Note: $N = 1129$. For each variable, the degrees of freedom (df) for the test is 2.

5.4 Additional Wald and LR Tests

5.4.1 Wald and LR Tests Based on a Single Reference Model

In recent years, the availability of programs for the estimation of nonparallel ordered regression models, such as the "vglm" and "rrvglm" functions from the VGAM package in R (Yee 2010) and the "gologit2" command in Stata (Williams 2006), has made it possible to test the parallel regression assumption with additional Wald and LR tests. Rather than estimating the binary cutpoint equations separately, as in the Brant test, it is now possible to estimate a nonparallel ordered regression model and to use this model for variable-specific and omnibus tests of the parallel regression assumption. If there are four outcome categories (1–4), then the null hypothesis for an omnibus Wald or LR test would be H_0: $\beta_{1k} = \beta_{2k} = \beta_{3k}$. For a model with five independent variables, the degrees of freedom for the test is $K(M - 2) = 5(4 - 2) = 10$. The LR test requires the estimation of both the nonparallel and parallel models, whereas the Wald test only requires the estimation of the nonparallel model. The main advantage of the Wald test over the LR test is that it is available for models using weighted data. Likelihood-based measures and tests, such as the LR test, Akaike information criterion (AIC), and Bayesian information criterion (BIC), do not work with survey weights because the assumption of independent observations no longer holds and the estimates are not based on a true likelihood (Long and Freese 2014, p. 102). The Wald test is also a more flexible tool for testing the equality of logit coefficients across multiple groups (Liao 2004).

For variable-specific tests, the Wald test once again uses the nonparallel model, but the LR test also requires the estimation of a partial model rather than a parallel model. Using the previous example, if we want to test the parallel regression assumption only for x_1, then we need to apply the following two constraints to the nonparallel model in order to produce the necessary partial model for the LR test: $\beta_{11} = \beta_{21}$ and $\beta_{11} = \beta_{31}$. In this partial model, the parallel regression assumption is retained for x_1 but relaxed for the remaining variables. The degrees of freedom for the Wald or LR test for x_1 is 2 because there are two constraints. We can test the parallel regression assumption for each variable separately by applying similar constraints and by estimating the appropriate partial models for the LR tests. For each LR test, the nonparallel model is the unconstrained model, and the partial model is the constrained model.

However, there is no reason that the nonparallel ordered model must be the reference model for tests of the parallel regression assumption. For variable-specific tests, the nonparallel model is a "reference" model in the sense that one compares the relative model fit of the nonparallel model to partial models, applying the parallel restriction to one variable at a time. If the nonparallel model provides a better fit, then this is evidence that the parallel assumption is violated for that variable. However, we also recommend using the parallel ordered model as the reference model. For a variable-specific test, we start with the parallel model and then relax the parallel restriction for one variable at a time. The comparison is between the parallel model and a partial model that applies the parallel restriction for every variable except one. If the partial model provides a better fit, then this is evidence that the parallel assumption is violated. The LR test requires the estimation of both models, whereas the Wald test only requires the estimation of the unconstrained model. Using either reference model, a significant Wald or LR test indicates that the parallel assumption is violated.* The choice of a parallel or nonparallel reference model is arbitrary. The nonparallel model may be preferable because it is potentially less biased than the parallel model, but it is also less parsimonious (Agresti 2010, p. 76). Therefore, we recommend conducting tests using both models.

For example, we may have a parsimonious model with only three independent variables: age, income, and education. Using a nonparallel reference model, we would test the parallel assumption for age by comparing the following models:

- Nonparallel: Relax the parallel assumption for all three variables
- Partial: Relax the parallel assumption for income and education but retain it for age

Alternatively, we could test the parallel assumption for age using a parallel reference model by comparing the following models:

- Parallel: Retain the parallel assumption for all three variables
- Partial: Relax the parallel assumption for age but retain it for income and education

Note that we do not use the same partial model in both tests.

In this example, it is not clear whether we should relax or retain the parallel assumption for income and education when we test the assumption for age. Therefore, we recommend testing the parallel assumption using both approaches. One should consider relaxing the parallel assumption if the test statistic is significant for at least one reference model. This discussion is not relevant for the omnibus test because it simply compares the parallel model to the nonparallel model. However, the results from variable-specific tests may vary depending on the reference model. The coefficients for a particular variable may vary more across

---

* Wald tests only require the estimation of the unconstrained model, which in this context is the model that retains the parallel regression assumption for the fewest number of variables. The nonparallel model is the unconstrained model using the nonparallel reference model, whereas the partial model is the unconstrained model using a parallel reference model.

cutpoint equations if the remaining coefficients are constrained because variation in the coefficients for one variable may be related to variation for other variables in the model.[*]

Another practical issue overlooked in discussions of the parallel regression assumption is whether the assumption holds for a subset of variables rather than for a single variable or the entire model. Researchers typically measure nominal-level independent variables with a set of binary indicators. In ordered regression models, one tests the joint significance of the group of binary variables with a Wald or LR test. We suggest taking the same approach with tests of the parallel regression assumption. For example, if there are two cutpoint equations and a set of three binary variables for region (South, Midwest, and West), then a joint test of the parallel regression assumption requires three constraints: $\beta_{1,south} = \beta_{2,south}$, $\beta_{1,midwest} = \beta_{2,midwest}$, and $\beta_{1,west} = \beta_{2,west}$. The logic of this test is also applicable to variables included in the model in quadratic form. For instance, in order to determine whether the nonlinear effect of age significantly varies across Cutpoint Equations 1 and 2, a test with the following two constraints is necessary: $\beta_{1,age} = \beta_{2,age}$ and $\beta_{1,age}^2 = \beta_{2,age}^2$.

5.4.2 Example #1: Cumulative Logit Model of General Self-Rated Health

In Chapter 2, we presented results from parallel cumulative models of general self-rated health. We have already presented evidence based on the score and Brant tests that the parallel regression assumption is violated for the entire model and for age, age-squared, and employed in particular. Results from the additional Wald and LR tests confirm these findings. In Table 5.2, we present the results from Wald and LR tests of the parallel assumption

TABLE 5.2

Results from Wald and LR Tests of the Parallel Regression Assumption for Parallel Cumulative Logit Models of General Self-Rated Health

	Reference = Nonparallel Model		Reference = Parallel Model	
	Wald	LR	Wald	LR
Omnibus test ($df = 20$)	37.09*	41.40**	37.09*	41.40**
Age and age-squared ($df = 4$)	9.22	11.60*	9.53*	12.83*
Age	8.76*	10.92**	5.40	5.44
Age-squared	8.11*	10.02**	3.72	3.56
Female	1.74	1.73	1.42	1.42
Race/ethnicity ($df = 6$)	3.50	3.39	5.84	6.36
Black	2.18	2.05	3.27	2.90
Latino	0.33	0.35	2.64	3.06
Other race/ethnicity	0.36	0.42	0.95	1.34
Education	1.32	1.28	0.48	0.48
Employed	14.64***	15.31***	17.72***	19.72***
Log income	0.12	0.12	1.17	1.14
Married	3.19	3.32	4.50	4.77

Note: For each variable, the degrees of freedom (df) is 2 for the Wald and LR tests.

*$p < 0.05$, **$p < 0.01$, ***$p < 0.001$.

---

[*] If the ordered regression model is set up using the person/threshold-level approach with threshold-specific observations for each respondent, then relaxing the parallel regression assumption is equivalent to adding interaction terms between the independent variable and the binary indicators for each cutpoint equation. The interaction terms may be more likely to be statistically significant if they are entered into the model 1 variable at a time.

using two different reference models. Using the nonparallel model, the effects of age, age-squared, and employed vary significantly across the three cutpoint equations. The omnibus test of the parallel regression assumption is also statistically significant. However, the results are somewhat different using the parallel model as the reference. The joint significance test for age and age-squared is significant using both the Wald and LR tests, but the separate tests for age and age-squared are nonsignificant.

Overall, the results in Table 5.2 suggest that it is reasonable to impose the parallel regression assumption for female, race/ethnicity, education, log income, and married. However, there is a fair amount of evidence that the parallel regression assumption is too restrictive for the effects of age, age-squared, and employed. The Wald and LR tests indicate that the coefficients significantly vary across cutpoint equations for these variables, but it is unclear whether this variation will translate into meaningful differences in the predicted probabilities.

5.4.3 Example #2: Continuation Ratio Logit Model of Educational Attainment

We also used Wald and LR tests in order to test the parallel regression assumption in the continuation ratio logit model of educational attainment introduced in Chapter 2 (see Table 5.3). The omnibus tests are significant, and the violation of the parallel assumption appears to be due to variation in the effects of age and parents' education across cutpoint equations. However, using the parallel reference model, the assumption is also violated for Latino. The inconsistent findings for Latino highlight the importance of considering both reference models. However, the joint tests are nonsignificant for race/ethnicity. One could also consider a number of different partial models as the reference, which is what iterative procedures do (see Section 5.4.5). Using a partial model with the parallel assumption relaxed for age and parents' education as the reference, the Wald and LR tests for Latino are not significant. On balance, the results suggest that it would be reasonable to retain the parallel assumption for the indicators of race and ethnicity.

TABLE 5.3

Results from Wald and LR Tests of the Parallel Regression Assumption for Parallel Continuation Ratio Logit Models of Educational Attainment

	Reference = Nonparallel Model		Reference = Parallel Model	
	Wald	LR	Wald	LR
Omnibus test ($df = 21$)	112.83***	118.05***	112.83***	118.05***
Age	10.26*	10.32*	27.38***	27.73***
Female	4.97	5.03	4.46	4.53
Race/ethnicity ($df = 9$)	10.81	11.11	15.71	15.69
Black	2.93	2.87	3.60	3.49
Latino	5.70	6.13	9.00*	9.08*
Other race/ethnicity	3.24	3.06	2.00	1.92
Parents' education ($df = 6$)	64.90***	64.86***	91.29***	92.50***
Mother's education	9.58*	9.72*	61.83***	63.41***
Father's education	26.87***	27.38***	77.73***	77.92***

Note: For each variable, the degrees of freedom (df) is 3 for the Wald and LR tests.

$*p < 0.05$, $**p < 0.01$, $***p < 0.001$.

TABLE 5.4

Results from Wald and LR Tests of the Parallel Regression
Assumption for Parallel Adjacent Category Logit Models
of Attitudes toward Welfare Spending

	Reference = Nonparallel Model		Reference = Parallel Model	
	Wald	LR	Wald	LR
Omnibus test ($df = 8$)	11.62	12.11	11.62	12.11
Age	2.67	2.68	2.04	2.05
Female	2.60	2.61	2.21	2.21
Education	5.84*	6.01*	4.94*	5.05*
Income	0.35	0.35	0.12	0.12
Ideology	1.41	1.42	0.10	0.10
Party identification ($df = 2$)	1.25	1.24	0.56	0.55
Democrat	1.22	1.21	0.43	0.43
Independent	0.89	0.88	0.07	0.07
Racial attitude	0.07	0.07	0.22	0.22

Note: For each variable, the degrees of freedom (df) is 1 for the
Wald and LR tests.
*$p < 0.05$, **$p < 0.01$, ***$p < 0.001$.

5.4.4 Example #3: Adjacent Category Logit Model of Welfare Attitudes

In our example for the adjacent approach, we introduced models of Whites' attitudes toward welfare spending in Chapter 2. According to the Wald and LR tests, the parallel regression assumption is appropriate for almost all of the variables (see Table 5.4). The omnibus Wald and LR tests are nonsignificant, which one could use to justify the choice of a parallel model. However, the variable-specific Wald and LR tests are significant for education using either reference model. Thus, a nonsignificant omnibus test does not guarantee that the parallel regression assumption will be reasonable for every variable in the model.

5.4.5 Iterative Wald and LR Tests

Researchers may use tests of the parallel regression assumption to select a particular ordered model. Nonsignificant test statistics for every independent variable suggest that the parallel model may be the best model; but if the test is significant for one or more variables, the researcher should consider using a partial or nonparallel model. One may use an iterative or stepwise procedure to find the "best" ordered regression model, although one should proceed cautiously using this approach. Starting with the nonparallel model, one would perform Wald or LR tests separately for each independent variable. For variables with nonsignificant test statistics, the parallel regression constraint would be applied first for the variable with the least significant test statistic (i.e., the largest p-value). The partial model with the parallel regression constraint imposed for one variable becomes the new reference model, and the process repeats until the Wald or LR tests are all statistically significant. Using a 0.05 alpha level, one would keep applying the parallel regression constraint to new variables through this iterative process until the Wald or LR tests all had p-values below 0.05 (Williams 2006).

We suggest proceeding cautiously with iterative Wald or LR tests because they require a larger number of tests for each variable; therefore, they increase the chances of over-fitting the data or "capitalizing on chance" (Williams 2006, p. 66). Scholars typically use the term over-fitting to refer to the estimation of too many parameters in a model. Iterative tests could lead one to choose a particular ordered regression model based on the peculiarities of the sample rather than on actual variation in coefficients across cutpoint equations in the population (Long 2015). Additionally, one is more likely to find a significant test statistic as the number of tests increases. Therefore, it would be reasonable to use a more conservative alpha level, such as 0.01, in order to avoid over-fitting the data. One could use a formal procedure for selecting a more conservative alpha level, such as the Bonferroni procedure (Aickin and Gensler 1996), which divides the alpha level by the number of tests. However, we rely on the 0.05 alpha level because formal tests are only a first step in the process of testing the parallel regression assumption. In our approach, these tests identify variables that should be the focus of subsequent tests based on predicted probabilities.

One may also avoid the problem of over-fitting the data in iterative tests by randomly splitting the sample into "exploration" and "verification" subsamples (Hastie et al. 2009; Long 2015). The iterative Wald and LR tests are based on the exploration subsample, and then the verification subsample is used to determine whether the selected model actually fits the data better than alternative models. If a partial model provides the best fit in the exploration sample, then the verification sample is used to compare its fit to parallel and nonparallel models. Alternatively, one could use the same iterative Wald or LR test in the verification subsample to make sure it selects the same model.[*]

5.5 Limitations of Formal Tests of the Parallel Assumption

The formal tests of the parallel regression assumption have several limitations. Therefore, researchers should not rely on the results of a single test statistic when evaluating the appropriateness of the parallel assumption for a particular variable or model. According to Peterson and Harrell (1990, p. 209), the LR test tends to perform the best, but the implementation of this test is "costly" in the sense that it requires the estimation of two ordered regression models. However, this cost is no longer a serious concern given the advances in statistical computing over the last 25 years. For an omnibus test of the parallel regression assumption, the LR test requires the estimation of the parallel and nonparallel ordered models. In the past, software restrictions made it difficult to estimate partial and nonparallel ordered regression models. Additionally, nonparallel models—such as the nonparallel cumulative logit model—are more likely to have convergence problems than the parallel cumulative logit model due to the potential for negative predicted probabilities (Agresti 2010; Yee 2015). The Wald test discussed in Section 5.4 suffers from the same problem as the LR test because it requires the estimation of a partial or nonparallel model. However, the score test only requires the estimation of the parallel model, and the Brant test is based on separately fitted binary regression models.

---

[*] We do not present examples of iterative Wald or LR tests in this book, but there are several examples available in an online appendix: http://sociology.okstate.edu/faculty-staff-directory/faculty-directory/dr-andrew-fullerton

Although the score test is easier to compute, it also has several well-known limitations. Models based on large sample sizes with many continuous independent variables are more likely to violate the parallel assumption according to the score test (Allison 1999, p. 141; O'Connell 2006, p. 47). In addition, a significant Brant test does not necessarily mean that the partial or nonparallel model should be preferred over the parallel model (Greene and Hensher 2010, p. 187). There are a number of factors that could lead to a significant Brant test statistic, including a misspecification of the linear predictor (e.g., omission of an important variable), heteroscedasticity, or the choice of a particular distributional form for the errors (e.g., logistic) (Brant 1990, p. 1173). Adding independent variables, using a heterogeneous choice model to account for heteroscedasticity (Williams 2009), or using a different link function may reduce or eliminate the coefficient variation across cutpoint equations. This actually applies to any formal test of the parallel assumption. Simply adding or removing a single independent variable can change the results of a test statistic.

Finally, traditional null hypothesis testing has been criticized because it may not be realistic to expect a coefficient or difference between two coefficients to be exactly zero (Gelman et al. 2012, p. 194). It is more reasonable to expect that the coefficients for a particular variable will be very similar but not exactly the same across cutpoint equations. This suggests that we should compare coefficient differences across cutpoint equations to a range of values (e.g., −0.1 to 0.1) rather than to zero. We will return to this point in our discussion of the Bayesian approach to ordered regression models in Chapter 6.

Due to these limitations, researchers should consider informal tests of the parallel regression assumption in addition to the formal tests. We will discuss several informal tests, including (1) model comparisons using information criterion measures, (2) comparisons of coefficients and odds ratios across equations in nonparallel models, and (3) comparisons of predicted probabilities and average marginal effects (AMEs) in parallel and nonparallel models. We will discuss all three types of informal tests but focus on examples of (1) and (3) due to the similarities between the latter two tests and the importance of predicted probabilities for determining practical or substantive significance.

5.6 Model Comparisons Using the AIC and the BIC

One may use information measures in order to informally test the parallel regression assumption. We will consider two information measures, which allow one to compare the relative fit of nested and nonnested models (see Weakliem 2004). The AIC is rooted in information theory (Akaike 1973). The AIC uses the likelihood as a measure of how well the model summarizes the data, but it adds a penalty based on the complexity of the model:

$$\text{AIC} = -2\text{LL} + 2k \tag{5.7}$$

where LL is the log likelihood, −2LL is the model deviance, and k is the number of free parameters in the model. The BIC is a measure rooted in Bayesian theory (Raftery 1995). It is also a model selection criterion based on the penalized log likelihood:

$$\text{BIC} = -2\text{LL} + (\ln[N])k \tag{5.8}$$

where N is the sample size.

For the AIC and BIC, smaller values correspond with a better model fit. Both information measures take the same general form: deviance + penalty. Without the penalty, the information measures would simply choose the most complex model as the "best" model. For each free or unconstrained parameter added to the model, the AIC imposes a penalty of 2 and the BIC imposes a penalty of $\ln(N)$. If $N = 1500$, then the BIC imposes a penalty of 7.3. In other words, the model deviance must decrease by more than 7.3 for each additional parameter in order for the more complex model to be a "better" model, according to the BIC (with a sample size of 1500). Although information measures such as the AIC and BIC are not statistical tests, researchers may use them in order to test the parallel regression assumption. For example, if one estimates parallel, partial, and nonparallel ordered regression models and the "best" model according to the AIC or BIC is one of the partial models, then this indicates that the parallel regression assumption is violated for a subset of the variables in the model.

The BIC tends to favor more parsimonious models than the AIC, which has important implications for the use of these information measures to test the parallel regression assumption. It is reasonable to use the AIC to test the parallel regression assumption for the entire model or a subset of variables because it penalizes the model based on the level of complexity, but it also tends to favor complex models if the added parameters help improve the model fit. One may also use the BIC to test the parallel regression assumption and in the process choose the "best" ordered regression model, but the BIC will tend to select very parsimonious models because it imposes a larger penalty than the AIC. In our experience, even if the parallel regression assumption is violated for a few variables in the model based on formal tests, the BIC will often indicate that the parallel model is the "best" model. As a result, relying solely on the BIC to test the parallel regression assumption may lead one to retain the parallel assumption even when there is a substantial amount of coefficient variation across equations. In this case, the model is "under-fitted" because there are not enough parameters in the model in order to account for the equation-specific effects of one or more variables.

5.6.1 Example #1: Cumulative Logit Model of General Self-Rated Health

The formal tests of the parallel regression assumption for the cumulative logit models of general self-rated health suggest that the assumption should be relaxed for age, age-squared, and employed. We present the AIC and BIC values for several cumulative logit models of general self-rated health in Table 5.5. Models 1 and 5 are the parallel and nonparallel models, respectively. Models 2 through 4 are partial models that relax the parallel assumption for different combinations of age, age-squared, and employed. For the parallel cumulative logit model, the AIC and BIC values are

$$AIC = 2519.770 + 2 \times 13 = 2545.770$$

$$BIC = 2519.770 + (\ln[1129]) \times 13 = 2611.149$$

The penalty for the model complexity is 26 for the AIC and 91.379 for the BIC.

The "best" model according to the AIC is the partial model that relaxes the parallel assumption for age, age-squared, and employed, whereas the "best" model according to the BIC is the partial model that only relaxes the parallel assumption for employed. The results clearly indicate that we should relax the assumption for employed, but there is also weaker evidence that we should relax the assumption for age and age-squared. Therefore, we should consider relaxing the parallel regression assumption for all three variables based on these information criteria.

TABLE 5.5

Tests of the Parallel Regression Assumption Based on Measures of Model Selection for Cumulative Logit Models of General Self-Rated Health

	Deviance	k	AIC	BIC
Model 1 (parallel)	2519.770	13	2545.770	2611.149
Model 2 (nonparallel effects for age and age-squared)	2506.940	17	2540.940	2626.434
Model 3 (nonparallel effects for employed)	2500.050	15	2530.051	2605.487
Model 4 (nonparallel effects for age, age-squared, & employed)	2489.494	19	2527.494	2623.047
Model 5 (nonparallel)	2478.366	33	2544.367	2710.327

Note: Deviance = −2(log likelihood) and k = number of free parameters.

TABLE 5.6

Tests of the Parallel Regression Assumption Based on Measures of Model Selection for Continuation Ratio Logit Models of Educational Attainment

	Deviance	k	AIC	BIC
Model 1 (parallel)	3312.032	11	3334.032	3390.784
Model 2 (nonparallel effects for age)	3284.302	14	3312.302	3384.532
Model 3 (nonparallel effects for parents' education)	3219.532	17	3253.532	3341.240
Model 4 (nonparallel effects for age and parents' education)	3210.272	20	3250.272	3353.457
Model 5 (nonparallel)	3193.980	32	3257.979	3423.077

Note: Deviance = −2(log likelihood) and k = number of free parameters.

5.6.2 Example #2: Continuation Ratio Logit Model of Educational Attainment

The Wald and LR tests of the parallel regression assumption in continuation ratio logit models of educational attainment indicate that we should relax the parallel regression assumption for age and the two measures of parents' education. In Table 5.6, we present the AIC and BIC for five continuation ratio logit models of educational attainment. In addition to the parallel and nonparallel models, there are three partial models that relax the parallel assumption for different combinations of age and parents' education. According to the AIC and BIC, the parallel regression assumption is too restrictive for the effects of parents' education. Both measures select models that relax the parallel assumption for these two variables. Relying on the BIC for model selection, we would choose the partial model that only relaxes the parallel assumption for parents' education, whereas the AIC would lead us to choose a partial model that also relaxed the assumption for age. On balance, these results suggest that we should seriously consider relaxing the parallel assumption for parents' education and possibly relax it for age as well.

TABLE 5.7

Tests of the Parallel Regression Assumption Based on Measures of Model Selection for Adjacent Category Logit Models of Attitudes Toward Welfare Spending

	Deviance	k	AIC	BIC
Model 1 (parallel)	733.581	10	753.581	793.294
Model 2 (nonparallel effects for age)	731.531	11	753.531	797.215
Model 3 (nonparallel effects for education)	728.527	11	750.527	794.211
Model 4 (nonparallel effects for age and education)	726.345	12	750.345	798.001
Model 5 (nonparallel)	721.476	18	757.476	828.959

Note: Deviance = −2(log likelihood) and k = number of free parameters.

5.6.3 Example #3: Adjacent Category Logit Model of Welfare Attitudes

For the adjacent category logit models of Whites' attitudes toward welfare spending, the Wald and LR tests were only statistically significant for education. However, in Table 5.7, we also consider partial models that relax the parallel assumption for age given the coefficient variation we observed for this variable in the nonparallel models. Once again, the model with the lowest BIC is more parsimonious than the model with the lowest AIC. Using the BIC, one would select the parallel model, whereas one would select the partial model with the assumption relaxed for age and education using the AIC. The AIC values are approximately the same for Models 3 and 4. Therefore, one could make the case that Model 3 is preferred because it is more parsimonious. The LR test comparing these two models is not statistically significant, which provides additional support for Model 3. Overall, these results reinforce the findings from the Wald and LR tests. The parallel regression assumption may be too restrictive for education, but we should retain the assumption for the remaining variables.

5.7 Comparing Coefficients across Cutpoint Equations

Another informal test of the parallel regression assumption is based on a comparison of coefficients, odds ratios, or hazard ratios across cutpoint equations in the nonparallel model (Kim 2003; Long 2015). If the coefficient or factor change score does not vary by a substantial amount across cutpoint equations in this informal test, then it is reasonable to use the parallel constraint even if the formal test is significant. The formal tests are based on the statistical significance of the differences across cutpoint equations, but this does not guarantee that the difference will be significant in a "practical" (Kim 2003) or substantive sense.

We could begin to examine the substantive significance of violations of the parallel regression assumption by focusing on variation in coefficients across cutpoint equations. This approach may be less useful for continuation ratio models because differences in unobserved heterogeneity across cutpoint equations are partially responsible for differences in the coefficients (Cameron and Heckman 1998). In order to avoid this problem, we can focus instead on differences in measures based on the predicted probabilities. However, these measures are potentially affected by sample selection bias in later stages, which is the main limitation in continuation ratio models.

5.8 Comparing AMEs and Predicted Probabilities across Models

The examination of differences in coefficients or odds/hazard ratios across cutpoint equations is a good informal test of the parallel regression assumption, but it is more important to consider differences in the predicted probabilities. Odds ratios can be misleading as an indicator of effect size because a large change in the odds does not always correspond with a large change in the predicted probability. The magnitude of the change in the predicted probability depends on the location of the change in x_k and on the locations of the remaining variables in the model. Therefore, we will examine the practical significance of tests of the parallel regression assumption using differences in the predicted probabilities and AMEs between the parallel and nonparallel models.

For the cumulative and adjacent models, one can examine differences in the AMEs and graphs of the predicted probabilities across the range of a continuous variable based on parallel and nonparallel models. A large difference in an AME or set of predicted probabilities between the two models is further evidence that one should relax the parallel regression assumption for that variable. For the stage models, one can examine differences in the conditional probabilities using AMEs and graphs as well.[*] One should use the conditional rather than unconditional predicted probabilities because the "total" effect on the unconditional predicted probability conflates changing effects across cutpoint equations with changes in the marginal outcome distribution across equations (Mare 1981).[†]

5.8.1 Example #1: Cumulative Logit Model of General Self-Rated Health

The parallel and nonparallel cumulative logit models of general self-rated health reveal important differences in the effects of age, age-squared, and employed on the predicted probabilities. The differences in AMEs among models are very substantial for employed (see Table 5.8). Employment has stronger effects on the predicted probabilities of poor and good health in the nonparallel model than in the parallel model. On average, being employed increases the predicted probability of good health by 0.10 in the nonparallel model but only by 0.02 in the parallel model. However, for other categories, such as excellent health, the AME is stronger in the parallel model.

There are also modest differences in the AMEs for age and age-squared. We examine the differences in the nonlinear effects of age using a graph of the predicted probabilities (see Figure 5.2).[‡] Relaxing the parallel regression assumption for age and age-squared does not have much effect on the predicted probabilities. The graphs for poor and fair health are

---

[*] The probabilities are conditional on making the transition to the current stage or cutpoint equation. For example, the conditional probability for Category 2 is the probability that $y = 2$ given that $y \geq 2$.

[†] One could also compare the "total marginal effects" (Buis 2015) in the parallel and nonparallel models.

[‡] For Figures 5.2 through 5.4, the predicted probabilities and 95% confidence intervals are based on the logit link with the remaining variables held constant at their means.

TABLE 5.8

Tests of the Parallel Regression Assumption Based on Differences in Average Marginal
Effects for Cumulative Logit Models of General Self-Rated Health

| | **(1 = Poor, 2 = Fair, 3 = Good, 4 = Excellent)** | | | | | | | |
| | **Parallel Model** | | | | **Nonparallel Model** | | | |
	$y = 1$	$y = 2$	$y = 3$	$y = 4$	$y = 1$	$y = 2$	$y = 3$	$y = 4$
Age	0.005***	0.012***	−0.001	−0.015***	0.012***	0.004	−0.004	−0.011*
	(4.038)	(4.521)	(−1.286)	(−4.531)	(4.061)	(0.801)	(−0.782)	(−2.403)
Age-squared[a]	−0.004***	−0.009***	0.001	0.012***	−0.010***	−0.002	0.004	0.008
	(−3.384)	(−3.653)	(1.257)	(3.661)	(−3.779)	(−0.574)	(0.837)	(1.681)
Employed	−0.033***	−0.088***	0.015*	0.106***	−0.070***	−0.078**	0.102**	0.046
	(−4.364)	(−4.703)	(2.105)	(5.141)	(−4.609)	(−2.731)	(3.073)	(1.664)

[a] Coefficient multiplied by 100 for ease of presentation. The numbers in parentheses are z-ratios.
*$p < 0.05$, **$p < 0.01$, ***$p < 0.001$.

very similar. The differences for excellent health are somewhat larger, but the overall age pattern is approximately the same in both models. Yet, the differences for good health are more important and may warrant relaxing the parallel regression assumption. There is an increase in the predicted probability of good health after the age of 60 that is only detectable by relaxing the parallel regression assumption.

The differences in the precision of the estimates are more substantial. There are similar patterns in the confidence intervals for fair and excellent health, but the patterns diverge for poor and good health. In the nonparallel model, the confidence interval for poor health gets wider from age 18 to 58 and then narrows from age 58 to 88. However, in the parallel model, the interval gets wider from age 18 to 58 and then remains relatively constant from age 58 to 88. For good health, the confidence interval is consistently wider in the nonparallel model and widest at age 88, whereas in the parallel model the interval is widest at age 18 and relatively constant from age 38 to 88. The probability of good health increases after age 60 in the nonparallel model, but the precision of the estimates decreases after that age as well.

Based on all of the formal and informal tests presented in this chapter, there is sufficient evidence to relax the parallel regression assumption for age, age-squared, and employed. However, the parallel assumption is empirically justified for the remaining variables. The formal tests were significant for age, age-squared, and employed, and the informal tests revealed important differences in the substantive significance of these effects between the parallel and nonparallel models. However, the substantive differences were larger for employment than for age.[*]

The nonparallel cumulative logit model reveals important insights that the parallel model overlooks. For example, employment has a stronger effect on the predicted probability of poor health than it does on excellent health. Nonparallel models allow asymmetrical patterns such as this to emerge by relaxing the parallel regression assumption. Having a job significantly decreases the probability of poor health, but it does not also increase the probability of excellent health. Based on the parallel cumulative model, one

---

[*] Additionally, we should note that informal tests based on graphs of predicted probabilities are sensitive to decisions regarding the values for the remaining independent variables in the model. However, we also compare AMEs, which are global measures that average the effects across every case in the sample.

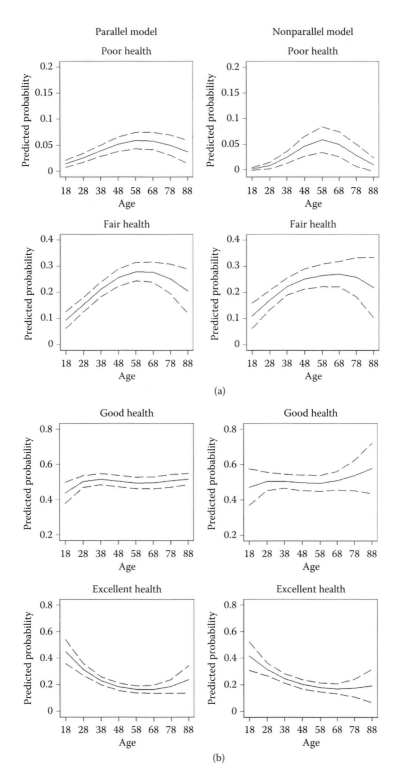

FIGURE 5.2
Relaxing the parallel assumption for age in cumulative logit models of general self-rated health.

would actually conclude that both effects are significant and the effect is stronger for excellent health than poor health.

The results of the formal and informal tests of the parallel assumption for models of health also highlight the usefulness of partial ordered regression models. Although the omnibus formal tests indicate that the parallel assumption is violated for the entire model, the variable-specific tests suggest that a partial model is preferred because the parallel assumption is reasonable for several variables in the model. Partial ordered regression models are a more parsimonious alternative to nonparallel models that researchers should consider using when the parallel assumption is only violated for a few variables in the model.

5.8.2 Example #2: Continuation Ratio Logit Model of Educational Attainment

In the continuation ratio logit models of educational attainment, there are several differences in the conditional AMEs for age and parents' education in the parallel and nonparallel models (see Table 5.9). For age, the conditional AMEs for associate and college are larger in the nonparallel model than in the parallel model. The AMEs are approximately the same for the conditional probability of a high school degree, but the significant, negative effect of age on the probability of attaining less than a high school degree is only present in the parallel model.

Relaxing the parallel regression assumption also has important implications for the effects of mother's and father's education. For mother's education, the conditional AMEs are smaller in the nonparallel models for the high school and college stages. For example, the average decrease in the conditional predicted probability of a college degree is almost twice as strong in the parallel model as it is in the nonparallel model (0.023 vs. 0.012). There are also fairly large differences in the effects of father's education on the conditional predicted probabilities in the high school and college stages. However, the effects of father's education are actually stronger for a high school degree in the nonparallel model. Additionally, relaxing the parallel assumption for father's education causes the effect on the conditional probability of a college degree to change direction. However, the effect in the final stage is relatively weak and only significant at the 0.10 level.

We examine the implications of relaxing the parallel regression assumption for mother's education in greater detail in Figure 5.3. There are substantial differences in the

TABLE 5.9

Tests of the Parallel Regression Assumption Based on Differences in Conditional Average Marginal Effects for Continuation Ratio Logit Models of Educational Attainment

	(1 = Less than High School, 2 = High School, 3 = Associate, 4 = College, 5 = Postgraduate)							
	Parallel Model				**Nonparallel Model**			
	$y=1$ if $y\geq1$	$y=2$ if $y\geq2$	$y=3$ if $y\geq3$	$y=4$ if $y\geq4$	$y=1$ if $y\geq1$	$y=2$ if $y\geq2$	$y=3$ if $y\geq3$	$y=4$ if $y\geq4$
Age	−0.001***	−0.003***	−0.002***	−0.002***	0.000	−0.003**	−0.004**	−0.003*
	(−3.982)	(−4.143)	(−4.004)	(−4.127)	(1.001)	(−2.975)	(−3.273)	(−1.961)
Mother's education	−0.010***	−0.027***	−0.018***	−0.023***	−0.014***	−0.019***	−0.018*	−0.012
	(−6.392)	(−7.028)	(−6.479)	(−6.891)	(−5.602)	(−3.374)	(−2.505)	(−1.370)
Father's education	−0.007***	−0.019***	−0.013***	−0.016***	−0.007***	−0.031***	−0.009	0.014
	(−5.555)	(−6.026)	(−5.556)	(−5.990)	(−3.339)	(−6.768)	(−1.568)	(1.877)

Note: The numbers in parentheses are z-ratios.
*$p < 0.05$, **$p < 0.01$, ***$p < 0.001$.

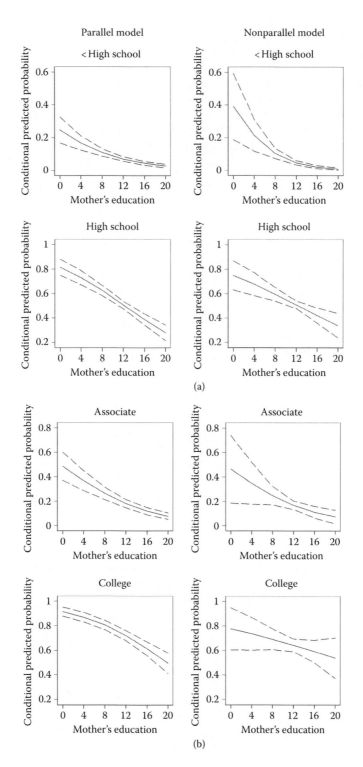

FIGURE 5.3
Relaxing the parallel assumption for mother's education in continuation ratio logit models of educational attainment.

conditional predicted probabilities for most educational levels between the parallel and nonparallel models. The effects of mother's education on the conditional probabilities for high school and college degrees are stronger in the parallel model, but the effect for less than a high school degree is stronger in the nonparallel model.

Relaxing the parallel regression assumption also has important implications for the precision of the estimates. In general, the confidence intervals are much larger at lower levels of mother's education in the nonparallel model. In the parallel model, the confidence interval for college is narrow and relatively constant across the range of mother's education. However, in the nonparallel model, the interval is largest at 0 years of mother's education, narrows substantially from 0 to 12 years, and then widens from 12 to 20 years. Based on these graphs, it is clear that the violation of the parallel regression assumption is both statistically and substantively significant for mother's education.

Overall, the results from the formal and informal tests of the parallel regression assumption for the model of educational attainment indicate that a partial model relaxing the parallel assumption for parents' education is preferred. Based on the parallel model, father's education reduces the probability of a college degree compared to a postgraduate degree, but relaxing the parallel assumption for this variable reveals a noticeably different pattern. Removing the parallel restriction, we see that father's education actually increases the predicted probability of a college degree, but the effect is only significant at the 0.10 level. The formal tests are also significant for age, but differences in the effects of age in the parallel and nonparallel models are relatively small for most educational levels.

5.8.3 Example #3: Adjacent Category Logit Model of Welfare Attitudes

Relaxing the parallel regression assumption has a noticeable impact on the AMEs for age and education in the adjacent category logit models of Whites' welfare spending attitudes (see Table 5.10). In the nonparallel model, there are stronger and statistically significant age differences in the predicted probabilities of opposition and mixed support.[*] In the parallel model, by contrast, the AMEs for age are nonsignificant. Education also has AMEs that are nonsignificant in the parallel model, but they are significant for one category in the

TABLE 5.10

Tests of the Parallel Regression Assumption Based on Differences in Average Marginal Effects for Adjacent Category Logit Models of Attitudes toward Welfare Spending

	(1 = Too Much, 2 = About Right, 3 = Too Little)					
	Parallel Model			**Nonparallel Model**		
	$y = 1$	$y = 2$	$y = 3$	$y = 1$	$y = 2$	$y = 3$
Age	−0.002	0.001	0.001	−0.003*	0.003*	0.000
	(−1.787)	(1.737)	(1.774)	(−2.352)	(2.064)	(0.440)
Education	0.003	−0.001	−0.002	−0.008	0.021*	−0.013
	(0.384)	(−0.384)	(−0.384)	(−0.867)	(2.292)	(−1.790)

Note: The numbers in parentheses are z-ratios.
*$p < 0.05$, **$p < 0.01$, ***$p < 0.001$.

---

[*] We refer to the welfare attitude categories as opposition ("too much"), mixed support ("about right"), and support ("too little").

nonparallel model. Education has a significant, positive association with the probability of mixed support for welfare spending in the nonparallel model. For both age and education, relaxing the parallel regression assumption leads to different substantive conclusions. Based on the parallel model, we would conclude that age and education are statistically unrelated to welfare attitudes. However, based on the nonparallel model, we would conclude that education and age have a significant, positive association with the probability of mixed support for welfare spending.

We examine the implications of relaxing the parallel regression assumption for education in more detail in Figure 5.4. Based on the graphs in Figure 5.4, it is clear that relaxing

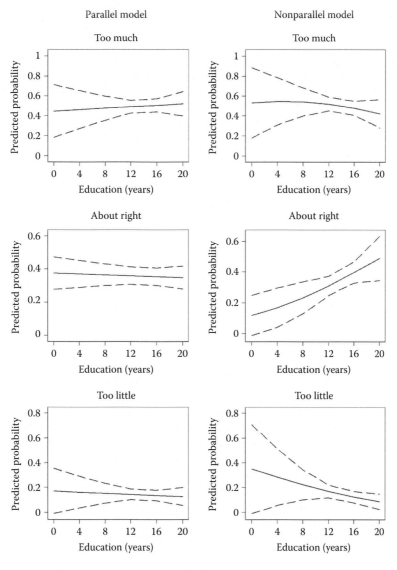

FIGURE 5.4
Relaxing the parallel assumption for education in adjacent category logit models of attitudes toward welfare spending.

the parallel assumption for education alters its effects on each category of welfare spending. Based on the parallel model, one would conclude that education has little to no association with welfare spending attitudes. However, in the nonparallel model, it is clear that education is positively associated with mixed support for welfare spending and negatively associated with both opposition and support. Additionally, education has much stronger associations with support and mixed support than with opposition to welfare spending. Across the range of education, the predicted probability of mixed support increases by 0.37, and the predicted probability of support decreases by 0.26. The predicted probability of opposition increases by 0.02 from 0 to 4 years of schooling and then decreases by 0.13 from 4 to 20 years of schooling. The patterns in the confidence intervals are similar in the parallel and nonparallel models, but the estimates are less precise in the nonparallel model.

Overall, the formal and informal tests of the parallel regression assumption indicate that we should relax the parallel regression assumption for education and consider relaxing it for age as well. Relying on the BIC, one would select the parallel adjacent category model, but every other formal or informal test of the parallel regression assumption suggests that a partial model without the parallel restriction for education is preferred over the parallel model. There is also a fair amount of evidence that we should relax the parallel assumption for age, including the AIC and the differences in the AMEs. The decision to retain or relax the assumption for age may come down to whether it is considered a key variable or simply a control variable in the model.

5.9 Conclusions

The parallel regression assumption constrains the coefficients in an ordered regression model so that they remain constant across cutpoint equations. This assumption results in more parsimonious models, but it also potentially produces misleading results and does not allow for asymmetrical effects. There are a variety of different tests of the parallel regression assumption. A robust examination of the parallel assumption requires the use of several different tests, including formal tests of statistical significance and informal tests based on model comparisons using information criterion and differences in predicted probabilities. No single test can determine whether the parallel assumption is reasonable for a particular variable or model. At a minimum, we recommend using one formal test, such as the LR test, and one informal test, such as a comparison of AMEs in the parallel and nonparallel models. A violation of the parallel regression assumption for one or more variables could indicate that the assumption of symmetry is not reasonable for that variable or group of variables. However, it could also indicate that the model is not properly specified or that the errors are heteroscedastic. The heterogeneous choice model explicitly incorporates heteroscedasticity into the ordered regression model (Williams 2009). We consider this model and other model extensions in the next chapter.

In the empirical examples, we saw that the results from formal tests of the parallel regression assumption vary depending on whether it is an omnibus or variable-specific test and a parallel or nonparallel reference model. A significant omnibus test is often the result of violations of the parallel assumption for only a few variables in the model, which is what we observed in the self-rated health and educational attainment examples. However, it is also possible to have a nonsignificant omnibus test but significant tests for one or more variables in the model. This is what we found for education in the adjacent category models of welfare attitudes. Based solely on the omnibus test, we would conclude that the parallel adjacent category model is preferred, but the application of the parallel regression assumption obscures the significant, positive effect of education on mixed support for welfare spending. The empirical examples also highlight the importance of combining formal tests of the parallel assumption with informal tests based on the AIC, BIC, AMEs, and on graphs of predicted probabilities. This combination of tests revealed that we should relax the parallel assumption for at least one variable in each example.

The trade-off between efficiency and bias in the population response proportions is important to consider when selecting an ordered regression model (Agresti 2010, p. 76). Parallel models are the most efficient because they take into account the order of the response categories and require the fewest number of parameters. However, they may also produce misleading results if the parallel regression assumption is violated for one or more variables. On the other hand, nonparallel models may have less bias, but they are also the least parsimonious. Additionally, the nonparallel adjacent category model does not take the category order into account given that it is equivalent to the multinomial logit model for nominal outcomes.

In general, we recommend using the model with the fewest parameters that yields approximately the same predicted probabilities as more complex models with additional parameters. If the predictions from a parallel ordered regression model are virtually the same as those from partial and nonparallel models, then the parallel model should be preferred because it maintains a stochastic ordering and requires the fewest parameters. For the cumulative approach, using the parallel model also allows one to avoid the problem of negative predicted probabilities inherent in partial and nonparallel models, which may cause convergence problems (Yee 2015). However, if the use of the parallel regression assumption leads to different conclusions about the direction of the relationship and statistical or substantive significance, then a partial or nonparallel model may be preferable over the parallel model. In addition, one should not use the similarity of the results from linear regression and parallel ordered regression models to justify the use of either model without testing the parallel regression assumption. The models may produce similarly misleading results if the parallel assumption is violated.

In this chapter, we compared differences in predictions among models using AMEs and graphs of predicted probabilities. However, one could also compare measures of model fit such as the ordinal pseudo-R^2 (Lacy 2006). In the three empirical examples, the nonparallel models explained an additional 0.9% to 1.5% of the ordinal variation compared to the parallel models, which provides additional support for the nonparallel models. The case for relaxing the parallel regression assumption is stronger if doing so leads to different substantive conclusions and a better model fit. Violations of the parallel regression assumption in formal tests do not necessarily mean that the substantive conclusions or model fit will change as a result of relaxing the assumption. Testing the parallel regression assumption is part of the larger process of choosing an ordered regression model, which requires the estimation of several alternative models. Fortunately, recent advances in statistical software applications have made this a relatively straightforward process (e.g., see Williams 2006; Yee 2010).

5.10 Appendix

5.10.1 Comparisons among the Wald, LR, and Score Tests

Buse (1982) proposed an effective graphical method to compare the Wald, LR, and score tests using the analogy of a hilltop (also see Engle 1984). Figure 5.5 is based on Buse (1982). In the graph, $\hat{\theta}_r$ and $\hat{\theta}$ are scalar ML estimates with and without constraints, respectively. The Wald test statistic measures the difference or horizontal distance between $\hat{\theta}_r$ and $\hat{\theta}$ adjusted by the variance of $\hat{\theta}$, which is the inverse of the negative expected value of the second derivative of the log-likelihood function with respect to θ evaluated at $\hat{\theta}$. According to Buse (1982, p. 154), when the second derivative is negative and large in magnitude, like the one for D_1 compared to that of D, then the distance between $\hat{\theta}_r$ and $\hat{\theta}$ is considered to be relatively large given the data, and one is more likely to reject the null. Despite its relatively high computational cost, the LR test simply compares the log likelihoods for the constrained and the unconstrained models, and the test statistic turns out to be proportional to the difference in the two log likelihoods.

Based on the rationale behind the score test, if the null hypothesis is true, then the gradient of the log likelihood with respect to the parameter estimates should not be significantly different from zero. In Figure 5.5, D and D_2 are two curves corresponding to different log-likelihood functions, or two different hilltops using Buse's analogy, that share the same tangent line, $g\left(\hat{\theta}_r\right)$, at point P corresponding to $\hat{\theta}_r$, the restricted ML estimate. Testing whether the gradient under the null is statistically different from zero requires additional information about the distribution of the log-likelihood function—namely, the curvature or the second derivative of the log-likelihood curve. Generally, if the second derivative is

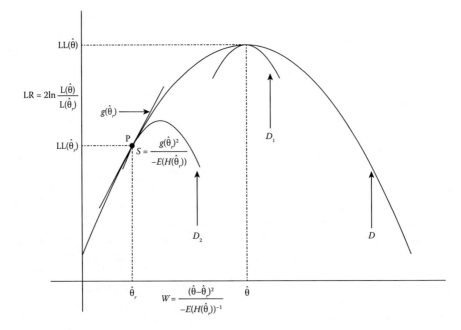

FIGURE 5.5
Likelihood ratio, Wald, and score tests (based on Buse, A. 1982. *American Statistician* 36: 153–157).

large and usually negative, then the inverse of the information matrix will be small and positive definite. This implies that the curvature or rate of change of the log-likelihood curve is large. If there is a point traveling along that parabola, D_2, from left to right, then that point travels a relatively short distance to the top of the hill. When this happens, the score test statistic is relatively small and it is unlikely to reject the null. Alternatively, if the second derivative is small, then the log-likelihood curve becomes flatter, and it is likely to reject the null because the score test statistic is relatively large. In Figure 5.5, it can be shown that the distance from P to the top of D_2, which has a relatively large curvature, is shorter than the distance to the top of D, which has a smaller curvature; thereby, $\hat{\theta}_r$ is closer to the unrestricted ML estimate for D_2 than it is to the unrestricted ML estimate for D and the null hypothesis with D_2 is less likely to be rejected than that with D.

Although the Wald, LR, and score tests provide asymptotically similar results, examinations of their empirical performance in finite samples have yielded useful guidelines for practitioners. The few studies that have adjudicated among the three tests (Engle 1984; Bera and McKenzie 1986; Agresti 2012; Tsai 2013) in general suggest that (1) notwithstanding its computational inconvenience, the LR test outperforms the other two, especially the Wald test, with its consistency and invariance properties in finite samples and (2) the score test is probably the optimal (i.e., most powerful given the same size) test under some general conditions. Engle (1984) also illustrates that, among the three, the score test is especially instrumental in providing model diagnostics with computational convenience. In practice, the Wald test is used often due to its programing convenience. In general, it is hard to say that one test is necessarily superior to the others under all circumstances. Each one has its advantages and disadvantages for different models and testing purposes.

6

Extensions

There are several ways to extend the ordered regression models considered thus far in order to account for more complex patterns and data structures. For example, researchers may have hypotheses that require the use of interaction terms, which is more complicated in categorical models such as ordered regression (Williams 2009; Mood 2010). The non-linear nature of the probability model can cause problems due to potential differences in unobserved heterogeneity among groups or levels of a variable included in the interaction term. One may address this problem by explicitly modeling the residual variance and the categorical outcome simultaneously with a "heterogeneous choice" or "location-scale" model (Clogg and Shihadeh 1994; Williams 2009). Additionally, the models in this book assume a single level of analysis, but more complex data structures with multiple levels of analysis are very common in the social and behavioral sciences, which requires the use of a multilevel ordered regression model (Raudenbush and Bryk 2002; Gelman and Hill 2007). Finally, advances in the Bayesian approach to statistics offer several advantages over the more familiar "frequentist approach" (Gelman et al. 2014; Gill 2015), including the ability to test multiple hypotheses simultaneously. This is particularly useful for tests of the parallel regression assumption, which require simultaneous tests for coefficients in multiple cutpoint equations. This chapter focuses on these three extensions to the ordered regression models in this book. To simplify the presentation, we will focus on the logit link function for each model extension.

6.1 Heterogeneous Choice Models

There are several factors that motivate statistical practitioners to estimate heterogeneous choice models instead of commonly used discrete choice models, such as the parallel cumulative logit model, for ordered responses. One is to further model the unexplained heterogeneities with re-parameterization, and a second is to relax the potentially unrealistic assumption of constant variance for the error term in the structural equation. The heterogeneous choice model is also a reasonable alternative to the nonparallel ordered regression model when the parallel regression assumption is violated, which may be the result of an inability to model the unobserved heterogeneity (Williams 2010). To derive the heterogeneous choice ordered response model, we may use the latent variable approach. Suppose there is a latent continuous variable, y^*, with the following structural model:

$$y^* = \mathbf{x}\boldsymbol{\beta} + \sigma\varepsilon$$

$$\tau_{m-1} < y^* < \tau_m \Leftrightarrow y = m \quad \forall y = 1, \ldots, m, \ldots, M$$

(6.1)

where $\mathbf{x}$ is a vector of independent variables, $\boldsymbol{\beta}$ is a column vector of structural parameters, ε is the error term that usually follows a standard normal or logistic distribution with a mean of zero, and σ is a positive scale factor. The scale factor adjusts the "true" residual variance so that it is equal to the assumed fixed value, such as $\pi^2/3$ in logit or 1 in probit. The heterogeneous choice model allows one to simultaneously model the ordinal outcome and the scale factor in order to account for differences in residual variance across cases. With some simple algebraic rearrangement, we have

$$\tau_{m-1} < \mathbf{x}\boldsymbol{\beta} + \sigma\varepsilon < \tau_m$$

$$\Rightarrow \frac{\tau_{m-1} - \mathbf{x}\boldsymbol{\beta}}{\sigma} < \varepsilon < \frac{\tau_m - \mathbf{x}\boldsymbol{\beta}}{\sigma} \tag{6.2}$$

The probability for a given category, m, is equal to the difference between two cumulative probabilities:

$$\Pr(y = m) = \Pr\left(\tau_{m-1} < y^* < \tau_m\right)$$

$$= F\left(\frac{\tau_m - \mathbf{x}\boldsymbol{\beta}}{\sigma}\right) - F\left(\frac{\tau_{m-1} - \mathbf{x}\boldsymbol{\beta}}{\sigma}\right) \tag{6.3}$$

If $\sigma = 1$ and the error term follows the standard logistic distribution, then Equation 6.3 reduces to the parallel cumulative logit model (Williams 2010, p. 545). If we allow σ to vary across individuals, thereby having heteroskedastic errors, then we have the heterogeneous choice model (Williams 2009). Because the scale factor, σ, is assumed to be positive, it is common to use an exponential function to parameterize it, such as $\sigma = \exp(\mathbf{w}\boldsymbol{\gamma})$, where $\mathbf{w}$ is a row vector of independent variables. There is at least one element in $\mathbf{w}$ that is different from $\mathbf{x}$, and $\boldsymbol{\gamma}$ is the corresponding column vector of parameters. Statistical software programs may estimate the log of the scale factor rather than the scale factor itself (e.g., see Williams 2010, p. 550).

6.2 Empirical Examples of Heterogeneous Choice Models

In Table 6.1, we compare results from the parallel cumulative logit and heterogeneous choice models using data from the 2012 GSS (Smith et al. 2013). The dependent variable is self-rated health, which is a four-category ordinal outcome ranging from excellent health (1) to poor health (4). For illustrative purposes, the two models include the most commonly used sociodemographic control variables, including age (in years), sex (male and female [ref.]), marital status (married and nonmarried [ref.]), race (Black, other, and White [ref.]), highest years of school completed (in years), log family income (log of imputed family income), and employment status (full-time, part-time, other, and unemployed [ref.]).

In general, the results suggest that the variance of the error term in the latent variable equation is associated with age and sex, providing support for the heterogeneous choice model. Relatedly, the magnitude of the effects in the heterogeneous choice model is greater than that in the classical parallel cumulative logit model. The Akaike information criterion (AIC) and Bayesian information criterion (BIC) lead to somewhat different

TABLE 6.1

Heterogeneous Choice and Parallel Cumulative Logit Models of Self-Rated Health

	Parallel Cumulative	Heterogeneous Choice
Age	0.017***	0.023***
	(0.003)	(0.005)
Sex (male = 1)	0.056	0.049
	(0.108)	(0.143)
Marital status (married = 1)	−0.292*	−0.384*
	(0.115)	(0.154)
Race (ref. = White)		
Black	0.050	0.060
	(0.148)	(0.190)
Other	0.037	0.046
	(0.176)	(0.225)
Years of education	−0.091***	−0.121***
	(0.019)	(0.027)
Log of family income	−0.234***	−0.309***
	(0.059)	(0.079)
Employment (ref. = unemployed)		
Full-time employed	−0.413	−0.543
	(0.251)	(0.327)
Part-time employed	−0.347	−0.463
	(0.280)	(0.360)
Other employment	−0.050	−0.087
	(0.254)	(0.327)
Thresholds		
Cutpoint 1	−4.220***	−5.548***
	(0.562)	(0.852)
Cutpoint 2	−2.050***	−2.738***
	(0.552)	(0.737)
Cutpoint 3	−0.192	−0.272
	(0.553)	(0.710)
ln(sigma): variance equation		
Age		0.004*
		(0.002)
Sex (male = 1)		0.182**
		(0.061)
LL	−1495.2145	−1488.482
AIC	3016.429	3006.964
BIC	3083.631	3084.504

Note: The numbers in parentheses are standard errors. $N = 1299$.
*$p < 0.05$, **$p < 0.01$, ***$p < 0.001$.

results, with the AIC favoring the heterogeneous choice model and the BIC not showing a strong preference for either model. In general, the BIC penalizes the number of parameters to a greater extent than the AIC does. However, some scholars argue that the AIC may have a sounder theoretical and empirical justification than the BIC (Burnham and Anderson 2004). The results from a likelihood ratio (LR) test also suggest that the heterogeneous choice model is preferred over the parallel cumulative logit model ($\chi^2 = 13.46$, $df = 2$, $p < 0.01$). We can use the same set of tools discussed in previous chapters to interpret the results from heterogeneous choice models, including odds ratios and average marginal effects. However, we also have to consider the residual variance equation in the interpretation.

In addition to the raw coefficients, we can examine the average predicted probabilities and marginal effects (see Table 6.2). The first four rows in Table 6.2 present the averaged predicted probabilities of having excellent (1), good (2), fair (3), and poor (4) health from the parallel cumulative logit and the heterogeneous choice models, respectively. An averaged predicted probability represents the predicted probability for a particular outcome category averaged across each case in the sample. The next four rows present average marginal effects on the predicted probability for each value of y with respect to years of education; that is, the marginal effect of education is averaged across each case in the sample. The results in Table 6.2 indicate that the averaged predicted probabilities and marginal effects for education are very similar in the two models. Supplementary analyses using hypothetical data points at a few extreme values, such as the maximum years of education, also yield substantively similar results for the two models. The similarity of the predictions provides support for the simpler, parallel cumulative logit model.

TABLE 6.2

Predicted Probabilities and Marginal Effects from Heterogeneous Choice and Parallel Cumulative Logit Models of Self-Rated Health

	Parallel Cumulative	Heterogeneous Choice
Averaged predicted probabilities ($y = 1$)	0.264*** (0.012)	0.265*** (0.012)
Averaged predicted probabilities ($y = 2$)	0.454*** (0.014)	0.455*** (0.014)
Averaged predicted probabilities ($y = 3$)	0.217*** (0.011)	0.215*** (0.011)
Averaged predicted probabilities ($y = 4$)	0.065*** (0.007)	0.065*** (0.007)
Average marginal effects for education ($y = 1$)	0.016*** (0.003)	0.017*** (0.003)
Average marginal effects for education ($y = 2$)	0.000 (0.001)	−0.000 (0.001)
Average marginal effects for education ($y = 3$)	−0.011*** (0.002)	−0.012*** (0.002)
Average marginal effects for education ($y = 4$)	−0.005*** (0.001)	−0.005*** (0.001)

Note: The numbers in parentheses are standard errors.
*$p < 0.05$, **$p < 0.01$, ***$p < 0.001$.

6.3 Group Comparisons Using Heterogeneous Choice Models

Heterogeneous choice models may also be useful when the goal is to compare effects among groups and the error variance is heteroskedastic. Such comparisons are traditionally executed in three ways (Xu and Fullerton 2014). One is to create an interaction term between the group indicator and a continuous variable and focus on the statistical significance of the interaction effect. A second option is to estimate the same regression model for each group and then compare the differences in the coefficients for the same independent variable among groups. This assumes that there is no correlation among the coefficients across the two subsamples. A third option is to delete the constant term in the structural model, include binary indicators of all subgroups (including the reference group), and to create interaction terms between these binary indicators and the variable of interest. In order to evaluate group differences in the effect of the continuous variable, one could conduct coefficient difference tests among the interaction terms because each one represents the effect of the variable of interest for a different subgroup.

The traditional ordered regression models assume that the error term follows a certain distribution and is homoscedastic. When this assumption is violated, structural parameter estimates tend to be biased and the results for group comparisons are simply incorrect (Long 2009).

For example, using the first method, if one hypothesizes that the effects of income on health are different for men and women, the structural model for the latent variable may be specified as

$$y_i^* = \sum Z\gamma + \beta x_i + \delta D_i + \lambda x_i D_i + \sigma_i \varepsilon_i$$

$$\Rightarrow \frac{y_i^*}{\sigma_i} = \sum \frac{Z\gamma}{\sigma_i} + \frac{\beta}{\sigma_i} x_i + \frac{\delta}{\sigma_i} D_i + \frac{\lambda}{\sigma_i} x_i D_i + \varepsilon_i$$

(6.4)

where Z is a vector of independent (control) variables, γ is a vector of parameters, x denotes the income variable, D is a binary group indicator for sex, and $x_i D_i$ is the interaction term; β, δ, and λ are regression population parameters; σ is a positive scale factor for the error term; and ε_i, the error term, is assumed to follow a particular probability distribution with a constant variance. If σ varies with certain values of the independent variables, then group comparison becomes implausible; that is, if we still follow the standard assumption of constant variance, then we will be estimating $\frac{\lambda}{\sigma_i}$ instead of λ as the population parameter for the interaction term. As a result, the significance test for group differences in the effect of income on self-rated health is invalid and misleading (Allison 1999; Williams 2009). Presumably, the heterogeneous choice model can resolve this issue by directly estimating σ_i.

To identify the model, one needs to assume that at least one slope is the same across groups when the heterogeneous choice model is used for binary outcomes. However, in ordered regression models, one may identify the model without this slope constraint by assuming that the cutpoints are the same across groups (Williams 2009; Mood 2010; Breen et al. 2014).

In general, the heterogeneous choice model can be quite useful in correctly estimating the structural parameters by explicitly modeling the heteroskedastic errors, and it is a potentially useful model when the parallel regression assumption is violated (Williams 2009; Agresti 2010). Table 6.3 presents the results from parallel cumulative logit and

TABLE 6.3

Heterogeneous Choice and Parallel Cumulative Logit
Models of Self-Rated Health with Interaction Terms

	Parallel Cumulative	Heterogeneous Choice
Age	0.017***	0.023***
	(0.003)	(0.005)
Sex (male = 1)	−1.929*	−2.453*
	(0.930)	(1.239)
Marital status (married = 1)	−0.284*	−0.376*
	(0.115)	(0.154)
Race (ref. = White)		
Black	0.046	0.056
	(0.148)	(0.190)
Other	0.037	0.048
	(0.176)	(0.225)
Years of education	−0.091***	−0.121***
	(0.019)	(0.028)
Log of family income	−0.318***	−0.399***
	(0.071)	(0.092)
Employment (ref. = unemployed)		
Full-time	−0.433	−0.569
	(0.252)	(0.327)
Part-time	−0.337	−0.455
	(0.280)	(0.361)
Other type	−0.052	−0.093
	(0.254)	(0.327)
Sex*income	0.201*	0.253*
	(0.093)	(0.124)
Thresholds		
Cutpoint 1	−5.051***	−6.445***
	(0.685)	(0.989)
Cutpoint 2	−2.876***	−3.624***
	(0.674)	(0.871)
Cutpoint 3	−1.016	−1.155
	(0.674)	(0.829)
ln(sigma): variance equation		
Age		0.004*
		(0.002)
Sex (male = 1)		0.176**
		(0.061)
LL	−1492.9042	−1486.3544
AIC	3013.808	3004.709
BIC	3086.179	3087.418

Note: The numbers in parentheses are standard errors. $N = 1299$.
*$p < 0.05$, **$p < 0.01$, ***$p < 0.001$.

heterogeneous choice models, both with an interaction term between sex and income. In the heterogeneous choice model, we also include an equation predicting $\ln(\sigma)$. Once again, the results suggest that the magnitude of the coefficients in the parallel cumulative logit model is more conservative. For example, the sex difference in the effect of income on the latent scale of self-rated health is 0.25 in the heterogeneous choice model but only 0.20 in the parallel cumulative logit model, which is a 20% difference. The two predictors in the heterogeneous equation are both statistically significant, which provides support for the heterogeneous choice model.

Table 6.3 also provides several statistics and measures for model comparisons. The BIC does not particularly prefer one model or the other because the difference in the BIC statistic is trivial between the two interaction models. However, the AIC suggests that the heterogeneous choice model with the interaction term is a better fitting model. This is also the preferred model according to the LR test ($\chi^2 = 13.1$, $df = 2$, $p < 0.01$).

Despite the usefulness of the heterogeneous choice model for ordinal models with interaction terms and more complex error structures, scholars have also noted a few limitations. Among the most important criticisms include the degree of inefficiency and bias associated with model misspecification and measurement errors in heterogeneous choice models (Keele and Park 2006). The heterogeneous choice model re-parameterizes the parallel cumulative model by also modeling a multiplicative scale factor, which allows one to predict the nonconstant error variance on the basis of one or more independent variables. In our experience, the results from heterogeneous choice models are often substantively similar to the results from simpler, parallel ordered models. Moreover, the application of heterogeneous choice models has typically been limited to parallel models due to the model's reliance on an error term in the structural model using a latent variable approach. However, heterogeneous choice models indirectly allow one to account for violations of the parallel regression assumption because nonconstant error variance is one potential explanation for this violation (Brant 1990, p. 1173). Therefore, researchers may consider using heterogeneous choice models in addition to partial and nonparallel models when the traditional parallel ordered model is inadequate (Williams 2010).

6.4 Introduction to Multilevel Ordered Response Regression

In the social and behavioral sciences, it is common for researchers to include variables measured at two or more levels of analysis in the same statistical model. For example, in studies of educational attainment variables may represent characteristics of students (Level 1) and schools (Level 2). The clustering of lower-level observations within higher levels and the inclusion of variables from multiple levels of analysis violates key assumptions of the traditional, single-level ordered regression model (e.g., independent and identically distributed errors). Therefore, one should use a multilevel ordered regression model that explicitly accounts for the multilevel nature of the data (Raudenbush and Bryk 2002; Gelman and Hill 2007; Rabe-Hesketh and Skrondal 2012). There is a structural model corresponding to each level with parameters from lower levels modeled as a function of parameters at higher levels. We can set up the multilevel ordered regression model by starting with the lower-level (Level 1) equation in a two-level model, without loss of generality.

Suppose we have an ordered response variable, $y = 1,\ldots, m,\ldots, M$, measured at the individual level, and $y = m$ when y^*, a latent, continuous variable, is set to be between two cutpoints, τ_{m-1} and τ_m, which are estimable from the data and the model. For the benchmark parallel cumulative logit model, in which ε follows a standard logistic distribution, we have

$$y_{ig}^* = x_{ig}\beta + \varepsilon_{ig}$$

$$P\left(y_{ig} = m \mid x\right)$$

$$= P\left(\tau_{m-1} < y_{ig}^* < \tau_m\right)$$

$$= F\left(\tau_m - x_{ig}\beta\right) - F\left(\tau_{m-1} - x_{ig}\beta\right) \qquad (6.5)$$

$$= \Lambda\left(\tau_m - x_{ig}\beta\right) - \Lambda\left(\tau_{m-1} - x_{ig}\beta\right)$$

$$\Rightarrow P\left(y_{ig} \le m \mid x\right) = \Lambda\left(\tau_m - x_{ig}\beta\right)$$

$$\Rightarrow \ln\left(\frac{P\left(y_{ig} \le m \mid x\right)}{1 - P\left(y_{ig} \le m \mid x\right)}\right) = \tau_m - x_{ig}\beta$$

where $\Lambda\left(\tau_m - x_{ig}\beta\right) = \dfrac{\exp\left(\tau_m - x_{ig}\beta\right)}{1 + \exp\left(\tau_m - x_{ig}\beta\right)}$, and subscripts i and g correspond to the lower-level and higher-level units, respectively (Hedeker and Gibbons 1994; Raudenbush and Bryk 2002; Rabe-Hesketh and Skrondal 2012). If we ignore the clustering of lower-level observations within higher-level units by treating the Level 1 coefficients as fixed, then the model reduces to a single-level ordered regression model. In the multilevel framework, one or more of the lower-level parameters vary across higher-level units. For example, if we are interested in the relationship between socioeconomic status (SES) and health for a population of individuals nested within communities, it is reasonable to hypothesize that some communities have a better benchmark health status to start with than others because of better air quality or self-selection, and it is likely that the relationship between SES and health may vary across communities as well (Kawachi et al. 2010).

The multilevel framework strikes a proper balance between model accuracy and efficiency by assuming that the key elements in the Level 1 coefficient vector follow certain probability distributions and may be conditional on one or more Level 2 characteristics. By taking the nested data structure into account, we may avoid problems such as downward bias in the standard errors for contextual effects and the inefficiency associated with single-level analyses at the macro level (Raudenbush and Bryk 2002).

6.4.1 Varying Intercepts

First, we will consider the multilevel parallel cumulative logit model with a random intercept, which allows the intercept or cutpoints in the parallel cumulative logit model to randomly vary across Level 2 units. Unless otherwise noted, we refer to random intercept models as "multilevel" models. Cutpoints and intercepts in ordered regression models are indistinguishable from each other, and they can be re-parameterized in various ways that are mathematically equivalent. In multilevel models, therefore, to have a varying intercept is equivalent to having varying cutpoints. The model may have an intercept, β_{0g}, that varies across Level 2 units, and each cutpoint can be expressed as a function of this intercept, $\tau_m = \beta_{0g} + \delta_m$. In order to estimate the cutpoints, it is necessary to make additional identifying assumptions. For example, it is common to assume that $\delta_1 = 0$ and $\tau_1 = \beta_{0g} \sim N(\gamma_{00}, \upsilon_{00})$. An alternative identifying assumption involves directly setting $\beta_{0g} \sim N(0, \upsilon_{00})$. Combining the equations from both levels, we have (Raudenbush and Bryk 2002; Rabe-Hesketh and Skrondal 2012)

$$\ln\left(\frac{P\left(y_{ig} \leq m \mid \mathbf{x}\right)}{1 - P\left(y_{ig} \leq m \mid \mathbf{x}\right)}\right) = \beta_{0g} + \delta_m - \mathbf{x}_{ig}\boldsymbol{\beta}. \tag{6.6}$$

To illustrate the applications of various multilevel ordered regression models, we use randomly sampled data from the 2007 International Social Survey Programme (ISSP) (ISSP Research Group 2009) with 2000 individuals nested within 33 countries. After list-wise deletion, our final sample size is 1830. The dependent variable is a five-category ordinal response measure of the frequency of physical activity ("take part in physical activities such as sports, going to the gym, and going for a walk"), ranging from never (1) to daily (5), with higher values corresponding to higher levels of physical activity. We also use a series of commonly used sociodemographic control variables as Level 1 predictors, including age (in years), family income (logged), sex (male = 1, female = 0), marital status (married = 1, other = 0), education (above higher secondary level, higher secondary level completed, and below higher secondary level [ref.]), employment status (employed, other employment, and unemployed [ref.]), and the type of city of residence (urban = 1, nonurban = 0).

The first column in Table 6.4 contains results from a multilevel parallel cumulative logit model using selected data from the 2007 ISSP. Most of the coefficients at Level 1 are in the expected directions. Having a high level of educational attainment, being married, and log income are associated with a higher level of self-rated physical activity, whereas the self-reported level of physical exercise decreases with age. The variance for the random intercept is 0.209 ($p < 0.05$), suggesting that there is a significant degree of clustering within countries. We can use the same methods of interpretation discussed in previous chapters to interpret the results from multilevel ordered regression models. However, we must also interpret the results for the random components, which refers to the variances and covariances of the random coefficients.

6.4.2 Varying Slopes

In multilevel models, Level 1 slopes can also vary across higher-level units. We may specify the structural model as follows:

$$\ln\left(\frac{P\left(y_{ig} \leq m \mid \mathbf{x}\right)}{1 - P\left(y_{ig} \leq m \mid \mathbf{x}\right)}\right) = \Lambda\left(\tau_m - \mathbf{x}_{ig}\boldsymbol{\beta}_g\right) \tag{6.7}$$

of which all or a subset of the coefficients in the $\boldsymbol{\beta}_g$ parameter vector vary across Level 2 units denoted by the subscript, g. The Level 2 model for a Level 1 slope takes the following form:

$$\beta_{kg} = \gamma_{k0} + \psi_{kg}, \; \psi_{kg} \sim N(0, v_k) \tag{6.8}$$

where $k \in [1, K]$ denotes a specific independent variable, g is the Level 2 unit, γ_{k0} is the population mean of β_{kg}, and ψ_{kg} is the error term that follows a normal distribution with

TABLE 6.4

Multilevel Parallel Cumulative Logit Models of Self-Rated
Physical Activity

	Random Intercept	Random Coef.
Age (in years)	−0.012***	−0.011***
	(0.003)	(0.003)
Sex (male = 1)	0.189*	0.187*
	(0.087)	(0.087)
Marital status (married = 1)	−0.061	−0.064
	(0.096)	(0.096)
Education (ref. = < higher secondary)		
Higher secondary level	0.238*	0.228
	(0.117)	(0.117)
More than higher secondary level	0.375**	0.370**
	(0.115)	(0.116)
Employment (ref. = unemployed)		
Other employment type	0.177	0.139
	(0.197)	(0.200)
Employed	−0.171	−0.201
	(0.191)	(0.194)
Locality (city = 1)	0.094	0.089
	(0.092)	(0.093)
Log of family income	0.475***	0.470***
	(0.058)	(0.071)
Thresholds		
Cutpoint 1	3.236***	3.115***
	(0.577)	(0.731)
Cutpoint 2	3.984***	3.868***
	(0.580)	(0.733)
Cutpoint 3	4.869***	4.753***
	(0.583)	(0.735)
Cutpoint 4	6.350***	6.231***
	(0.588)	(0.739)
Variance component		
Intercept	0.209***	4.680
	(0.071)	(3.828)
Log imputed income in US dollars		0.036
		(0.034)
Covariance (intercept, income)		−0.403
		(0.362)
LL	−2714.1072	−2716.2156
AIC	5460.431	5460.214
BIC	5537.600	5548.407

Note: The numbers in parentheses are standard errors. The variance components are reported in standard deviations. $N = 1830$.

*$p < 0.05$, **$p < 0.01$, ***$p < 0.001$.

a mean of zero and variance of v_k (Hedeker and Gibbons 1994; Raudenbush and Bryk 2002; Rabe-Hesketh and Skrondal 2012).

The second column in Table 6.4 presents results from a multilevel random slope model using the same data from the previous example. In this model, both the intercept and the slope for logged family income are allowed to vary across countries, and the Level 2 errors are unstructured (i.e., the random coefficients are allowed to be correlated). Results for the regression parameter estimates at Level 1 are very similar to those from the random intercept model, except that the magnitude and significance level of parameter estimates appear to be smaller for each variable. This is not surprising because as we inflate the variance in the error term as we did in the heterogeneous choice models, the effects of independent variables are suppressed by design. However, none of the random components are statistically significant. We also estimated a second random slope model while constraining the covariance between the intercept and the slope for logged family income to be zero. Results (not shown) indicate that the variance for the intercept, but not the logged family income, is still statistically significant. This provides a good example of how sensitive multilevel ordered regression models are to model specification. In addition, model convergence is more difficult as the number of random coefficients increases, particularly when the sample size at Level 2 is relatively small.

6.4.3 Intercepts and Slopes as Outcomes

In the multilevel modeling framework, both intercepts and slopes in lower-level equations may be modeled or predicted by Level 2 variables. For example, it is quite plausible to hypothesize that neighborhood characteristics, such as the average income or education in a community, can enhance or exacerbate the benchmark health status or moderate the relationship between income and health at the individual level. If the health literature or qualitative evidence has suggested such a possibility, then we can add variables to the Level 2 equations:

$$\beta_{0g} = \gamma_{00} + \sum_{h=1}^{H_0} \gamma_{0h} Z_{hg} + \psi_{0g}$$

$$\beta_{kg} = \gamma_{k0} + \sum_{h=1}^{H_k} \gamma_{kh} Z_{hg} + \psi_{kg} \qquad \Psi \sim N(0, \mathbf{N})$$

$$(6.9)$$

where Ψ is a $1 \times (K + 1)$ vector containing ψ_{kg}, and $\mathbf{N}$ is a $(K + 1) \times (K + 1)$ symmetric positive definite matrix with its diagonal elements containing the variance components for ψ_{kg} and off-diagonal elements containing the covariance between ψ_{kg} and ψ_{jg} for $j \neq k$ (Raudenbush and Bryk 2002; Rabe-Hesketh and Skrondal 2012). We estimate the main effects of Level 2 variables in the intercept equation and cross-level interactions in the slope equations.

To demonstrate, we use the same five-category ordered response variable for physical activity from the 2007 ISSP data. Building on conventional wisdom in this line of research, we use gross domestic income per capita in 2007 as our Level 2 predictor. We assume that this variable can help explain the variation in the intercept and the income coefficient at Level 1. Table 6.5 presents the results from this multilevel model.

TABLE 6.5

Slopes-as-Outcomes Multilevel Parallel Cumulative
Logit Model of Self-Rated Physical Activity

Level 1	
Age (in years)	−0.015***
	(0.003)
Sex (male = 1)	0.196*
	(0.091)
Marital status (married = 1)	−0.032
	(0.101)
Education (ref. = < higher secondary)	
Higher secondary level	0.263*
	(0.122)
More than higher secondary level	0.389**
	(0.121)
Employment (ref. = unemployed)	
Other employment type	0.094
	(0.200)
Employed	0.171
	(0.196)
Locality (city = 1)	0.043
	(0.098)
Thresholds	
Cutpoint 2 δ_2	0.772***
	(0.046)
Cutpoint 3 δ_3	1.683***
	(0.063)
Cutpoint 4 δ_4	3.193***
	(0.089)
Level 2	
Cutpoint 1	
Intercept	−4.662***
	(1.020)
GDP per capita (in 1k)	0.097**
	(0.035)
Log of family income	
Intercept	0.618***
	(0.107)
GDP per capita (in 1k)	−0.008***
	(0.003)

Note: GDP = gross domestic product. The numbers in parentheses are standard errors. The variance components are reported in standard deviations. $N = 1799$.
*$p < 0.05$, **$p < 0.01$, ***$p < 0.001$.

According to the results, being male, having higher educational attainment, and income are associated with an elevated level of physical activity. Results from Level 2 provide evidence for the relevance of economic development to a healthy lifestyle. As gross domestic product (GDP) per capita increases, average physical activity also is predicted to increase when income = 0. This effect is conditional on income given the inclusion of the cross-level interaction between GDP and income. According to this cross-level interaction effect, GDP per capita suppresses the positive effects of family income on physical activity, leading to a higher level of lifestyle parity across different income levels. However, interaction effects are more complicated in ordered regression models due to the nonlinear nature of the probability models. Therefore, researchers should be more cautious in the use of cross-level interactions in multilevel ordered regression models.

6.4.4 Multilevel Nonparallel Ordered Regression Models

The parallel regression assumption is just as important in multilevel ordered regression models as it is in single-level models, but researchers often overlook it due to the increased model complexity and lack of tests for the parallel assumption in standard multilevel software packages. The traditional multilevel ordered regression model requires the parallel regression assumption for every variable in the model. In a multilevel nonparallel cumulative logit model, the slopes are allowed to vary across equations, and the intercept or cutpoints vary across higher-level units:

$$\ln\left(\frac{P(y_{ig} \le m \mid x)}{1 - P(y_{ig} \le m \mid x)}\right) = \Lambda\left(\beta_{0g} + \delta_m - x_{ig}\beta_m\right) \tag{6.10}$$

In Equation 6.10, the intercept, β_{0g}, is set to vary across Level 2 units. We can also allow one or more slopes to vary across higher-level units:

$$\beta_{mkg} = \gamma_{mk0} + \psi_{mkg} \qquad \Psi \sim N(0, \mathbf{N}) \tag{6.11}$$

or

$$\beta_{mkg} = \gamma_{mk0} + \sum_{h=1}^{H_{mk}} \gamma_{mkh} Z_{mhg} + \psi_{mkg} \qquad \Psi \sim N(0, \mathbf{N}) \tag{6.12}$$

by having Level 2 predictors for β_{mkg} (Rabe-Hesketh and Skrondal 2012). However, as more random effects are added, model convergence tends to become an issue.

To illustrate, we will consider the example of physical activity from the 2007 ISSP. The parallel regression assumption is relaxed for every variable in the model, and the intercept is allowed to randomly vary across countries. The results from the multilevel nonparallel cumulative logit model in Table 6.6 in large part agree with those from previous multilevel models that education and income have positive effects on physical activity, whereas age is negatively related to it. In addition, the predictors seem to differentiate the odds between never/rarely versus a higher frequency of physical activity more so than the odds of the most frequent (daily) versus other. This is an informal indication of the violation of

TABLE 6.6

Multilevel Nonparallel Cumulative Logit Model of Self-Rated Physical Activity

	1 vs. 2/5	1/2 vs. 3/5	1/3 vs. 4/5	1/4 vs. 5
Age (in years)	0.018***	0.012***	0.008*	0.004
	(0.004)	(0.003)	(0.003)	(0.005)
Sex (male = 1)	−0.370**	−0.265*	−0.056	0.008
	(0.118)	(0.104)	(0.102)	(0.139)
Marital status (married = 1)	0.020	0.234*	0.132	−0.130
	(0.120)	(0.112)	(0.112)	(0.154)
Education (ref. = < higher secondary level)				
Higher secondary level	−0.355*	−0.283*	−0.041	−0.170
	(0.148)	(0.135)	(0.137)	(0.184)
More than higher secondary level	−0.546***	−0.487***	−0.335*	−0.197
	(0.158)	(0.136)	(0.132)	(0.174)
Employment (ref. = unemployed)				
Other employment type	−0.196	−0.191	−0.111	−0.135
	(0.232)	(0.223)	(0.229)	(0.295)
Employed	−0.142	0.119	0.326	0.483
	(0.226)	(0.217)	(0.223)	(0.292)
Locality (city = 1)	0.083	−0.086	−0.126	−0.186
	(0.118)	(0.108)	(0.107)	(0.142)
Log of family income	−0.674***	−0.521***	−0.374***	−0.147
	(0.067)	(0.064)	(0.064)	(0.076)
Constant	5.026***	4.367***	3.732***	3.147***
	(0.651)	(0.627)	(0.635)	(0.756)
Variance components country	0.473***			
	(0.081)			
AIC	5351.770			
BIC	5577.764			

Note: The numbers in parentheses are standard errors. $N = 1830$.
*$p < 0.05$, **$p < 0.01$, ***$p < 0.001$.

the parallel regression assumption. An LR test comparing the multilevel parallel and non-parallel cumulative logit models also indicates that the parallel assumption is violated for the entire model ($\chi^2 = 162.66$, $p < 0.01$).

6.4.5 Multilevel Ordered Regression Models Using Alternative Link Functions

The previous sections on multilevel ordered response models focus solely on the cumulative logit link, in which the error term in the structural model for the latent variable is assumed to follow a standard logistic distribution. However, if other link functions or modeling strategies are used, or parameter constraints are imposed, then one could consider several alternative multilevel ordered regression models. The number of potential models can be quite large ($3 \times 3 \times 3 = 27$) if one simply multiplies the ways to categorize ordered response models (Fullerton 2009). It is almost impossible to include all of these types under the multilevel modeling framework; nor are these models all substantively meaningful in most empirical scenarios. Although some statistical packages may have a few of these variants and extensions, none can claim to cover every model.

The challenges often stem from difficulties in numerical estimation. The use of Bayesian estimation and inference, however, has become a feasible and attractive alternative with current computational power for simulation-based methods.

6.5 Bayesian Analysis of Ordered Response Regression

Bayesian statistics sets up a different framework than the traditional "frequentist" approach, which typically views probability as the result of long-run trials and relies on the hypothetical notion of repeated sampling. Rather than insisting on an absolute, fixed state of reality, Bayesian inference allows for different "degrees of belief" or "Bayesian probabilities" (Kruschke 2015), which in many cases is interpreted as "subjective probability." It is based on Bayes' theorem on conditional probabilities:

$$P(A \mid B) = \frac{P(B \mid A)P(A)}{P(B)} \tag{6.13}$$

which can be easily converted to

$$P(\theta \mid D) = \frac{P(D \mid \theta)P(\theta)}{P(D)} \tag{6.14}$$

if A and B in Equation 6.13 are supplanted with θ and D, respectively. Here θ usually corresponds to population parameters and D to observed data (Agresti and Hitchcock 2005; Gelman and Hill 2007; Carlin and Louis 2009; Agresti 2010; Lynch 2010; Kruschke 2013; Gelman et al. 2014; Gill 2015; Kruschke 2015). According to Laplace (1986, pp. 364–365), this suggests that the probability of a possible "cause" or parameter given the "effect" or data are proportional to the probability of observing the same data that are generated by the parameters of interest. This seemingly simplistic but insightful result has been used in human interactions and many decision-making processes.

According to Bayesian inference, $P(\theta \mid D)$ is the joint posterior probability distribution of parameter θ, $P(D \mid \theta)$ is the likelihood (i.e., the probability of observing the data given the parameters), and $P(\theta)$ is the prior probability distribution of θ, which usually contains our preconception about how θ is distributed. Because D (data) is usually seen as fixed and $P(D)$ is a marginal distribution with θ being integrated out, $P(D)$ is usually treated as a normalizing factor. Because p usually denotes the probability mass function in statistics, we may express the joint posterior as

$$p(\theta \mid D) \propto p(D \mid \theta)\, p(\theta) \tag{6.15}$$

indicating that the joint posterior is proportional to the product of the likelihood and the prior probability (Gelman et al. 2014; Kruschke 2015). Conceptually, this makes Bayesian estimation seemingly quite straightforward; that is, once the prior and the likelihood are specified, we can derive the posterior by simply carrying through the multiplication. When one has very limited knowledge about the prior, then a noninformative prior may be used, which tends to have a large variance. No preferences are given in the sampling parameters in a specific region of the probability distribution. When a noninformative prior is used together with a large sample, the results from Bayesian and frequentist models tend to be very similar.

In deriving the posterior, the product of the likelihood and the prior may be a commonly recognizable probability distribution function. For example, if one uses a conjugate prior, which is a prior that yields a posterior from the same distributional family, then the computation can be made quite straightforward and results are easily interpretable using sufficient statistics (Gelman et al. 2014). However, not every empirical problem can be dealt with using conjugate priors. In many cases, the product of the likelihood function and the prior could be an unrecognizable function or even an improper distributional function. Analytical solutions are often not available. In that case, $p(\theta|D)$ may be obtained using the Markov Chain Monte Carlo (MCMC) simulation method (see Gelman et al. 2014) once the prior and likelihood are properly supplied (for more details, see Kruschke 2015, pp. 117–177).

Bayesian estimation and inference can perform surprisingly well in cases where the traditional frequentist approach would usually fail. Computational difficulties are common in both Bayesian and frequentist models when the models are complex or the likelihood function has a "plateau." With the choice of a proper prior, however, Bayesian estimation can sometimes circumvent problems such as underidentification and singularities in matrix inversion. In addition, Bayesian inference provides the probability that a hypothesis is true given the data. Some researchers mistakenly associate this interpretation with the frequentist p-value, which is actually the probability of the data given that the null hypothesis is true. By having a multivariate posterior, Bayesian inference also avoids the often imprecise and cumbersome adjustments in simultaneous hypothesis testing under the frequentist framework (Gelman et al. 2012; Kruschke 2013). Despite these advantages, some scholars have reservations about fully implementing Bayesian inference in practice because it comes with the relatively high computational cost, it requires a more extensive background in mathematics, and the choice of priors is inherently subjective and therefore subject to criticism (Gelman and Robert 2013a, 2013b).

6.6 Empirical Examples of Bayesian Ordered Regression Models

In the following examples, we illustrate the Bayesian estimation of parallel cumulative logit, nonparallel cumulative logit, and multilevel stereotype logit models with noninformative priors. The sample size is relatively large in each example as well. Therefore, the results should be similar for Bayesian and frequentist models. However, the Bayesian models provide more meaningful ways to interpret certain quantities, such as p-values, and make it easier to simultaneously test multiple hypotheses.

6.6.1 Bayesian Estimation of the Parallel Cumulative Model

In the first empirical example, we will focus on the Bayesian estimation of a parallel cumulative logit model using the self-rated health example from the 2012 GSS. Using Bayes' theorem, the joint posterior probability distribution of the population parameters, β and τ, is

$$p(\beta, \tau|y) \propto p(y|\beta, \tau)\, p(\beta, \tau) \tag{6.16}$$

in which

$$p(y \mid \beta, \tau) = \prod_{i=1}^{N} \left(F(\tau_m - x_i\beta) - F(\tau_{m-1}x_i\beta) \right)$$
(6.17)

$$p(\beta, \tau) = p(\beta)p(\tau)$$

Noninformative priors can be chosen for $p(\beta)$ and $p(\tau)$ as follows:

$$p(\beta) \sim N(0, 1.0E - 6) \forall \beta$$
$$p(\tau_m) \sim N(m + 0.5, 1.0E - 6) \forall \tau_m$$
(6.18)

For Bayesian estimation of the model, we used JAGS 3.4.0 through rjags in R 3.0.2 (Plummer 2003; R Core Team 2014). In many software applications, users are requested to specify the inverse of the variance (i.e., the precision parameter) for a normal distribution; m in (6.18) denotes the response value for y, and setting the mean for τ_m to $m + 0.5$ is necessary to retain the ordinality among cutpoints. Three chains were used with 2000 burn-in (Gelman et al. 2014, p. 281) and adaptation (Gelman et al. 2014, p. 308) steps, respectively—which are usually required during the preparatory stage for a MCMC sampling process—and there were a total of 50,000 iterations to reach convergence.

Table 6.7 presents the results from parallel cumulative logit models of self-rated health from the 2012 GSS estimated using the Bayesian and frequentist (i.e., maximum likelihood estimation, or MLE) approaches. Unlike MLE, Bayesian estimation provides a single joint posterior distribution and summary statistics for that joint distribution. The first column in Table 6.7 contains the means of structural (β) and auxiliary (τ) parameters from the posterior; the credible intervals for those parameters are in parentheses, which provide the bounds for the middle portion of the posterior with equal tails. For example, a 95% credible interval gives the lower and upper bounds for the middle 95% of the posterior distribution. Unlike the precision estimates in the frequentist framework, we can comfortably make probabilistic statements about the uncertainty associated with the parameters because the posterior is a probability distribution. For example, we can state that there is a 0.95 probability that the parameter for age falls between 0.010 and 0.023. The posterior distributions of other quantities of our interest—such as odds ratios, predicted probabilities and marginal effects—can be readily derived from the posteriors of these raw coefficients using well-established mathematical relationships. Note that the results from our Bayesian estimation and those from MLE are quite similar and even identical to the third decimal place in some cases due to the use of noninformative priors.

6.6.2 Bayesian Estimation of the Nonparallel Cumulative Model

Bayesian estimation of the nonparallel cumulative logit model is also straightforward if we use noninformative priors. The joint posterior is almost exactly the same as that in (6.16), except that we have a separate β_m for each cumulative logit comparison of the sufficient set. The likelihood becomes $\prod_{i=1}^{N} \left(F(\tau_m - x_i\beta_m) - F(\tau_{m-1}x_i\beta_m) \right)$ and the priors for β_m's are now $p(\beta_m) \sim N(0, 1.0E - 6) \forall \beta_m$.

Table 6.8 presents results from a Bayesian estimation of a nonparallel cumulative logit model. Testing the parallel regression assumption in ordered regression models is more

TABLE 6.7

Bayesian and Maximum Likelihood Estimation of Parallel Cumulative
Logit Models of Self-Rated Health

	Bayesian Estimation $\bar{\beta}$ (95% Credible Interval)	MLE $\hat{\beta}$ (95% Confidence Interval)
Age	0.017	0.017***
	(0.010, 0.023)	(0.010, 0.023)
Sex (male = 1)	0.056	0.056
	(−0.157, 0.270)	(−0.156, 0.268)
Marital status (married = 1)	−0.298	−0.292*
	(−0.519, −0.074)	(−0.518, −0.067)
Race (ref. = white)		
Black	0.053	0.050
	(−0.236, 0.340)	(−0.240, 0.340)
Other	0.038	0.037
	(−0.302, 0.380)	(−0.309, 0.383)
Years of education	−0.093	−0.091***
	(−0.131, −0.055)	(−0.129, −0.053)
Log of family income	−0.228	−0.234***
	(−0.337, −0.112)	(−0.349, −0.118)
Employment (ref. = unemployed)		
Full-time employed	−0.411	−0.413
	(−0.890, 0.090)	(−0.906, 0.080)
Part-time employed	−0.342	−0.347
	(−0.884, 0.207)	(−0.895, 0.201)
Other employment	−0.043	−0.050
	(−0.532, 0.460)	(−0.547, 0.447)
Thresholds		
Cutpoint 1	−4.186	−4.220***
	(−5.171, −3.143)	(−5.321, −3.118)
Cutpoint 2	−2.008	−2.050***
	(−2.972, −0.978)	(−3.131, −0.968)
Cutpoint 3	−0.140	−0.192
	(−1.105, 0.895)	(−1.276, 0.892)
LL		−1495.2145
AIC		3016.429
BIC		3083.631
DIC	3021.430	

Note: The numbers in parentheses are standard errors. $N = 1299$.
*$p < 0.05$, **$p < 0.01$, ***$p < 0.001$.

complicated than many researchers may realize. One major challenge that comes with the
traditional frequentist approach is related to the necessity of multiple comparisons; that is,
one needs to make an adjustment, such as the Bonferroni correction, to the p-value or the
level of α because tests of the parallel assumption require the simultaneous testing of several null hypotheses related to coefficients in each cutpoint equation. For example, to test
the parallel assumption in a nonparallel ordered regression model with a four-category

TABLE 6.8

Bayesian Estimation of a Nonparallel Cumulative Logit Model of Self-Rated Health

	2/4 vs. 1 β (95% Credible Interval)	3/4 vs. 1/2 β̄ (95% Credible Interval)	4 vs. 1/3 β̄ (95% Credible Interval)
Age (in years)	0.014	0.019	0.012
	(0.006, 0.023)	(0.011, 0.027)	(−0.002, 0.026)
Sex (male = 1)	0.122	0.185	0.587
	(−0.384, 0.140)	(−0.084, 0.445)	(0.120, 1.061)
Marital status (married = 1)	−0.423	−0.198	−0.224
	(−0.706, −0.143)	(−0.478, 0.085)	(−0.732, 0.274)
Race (ref. = White)			
Black	0.250	0.043	−0.804
	(−0.134, 0.651)	(−0.322, 0.396)	(−1.607, −0.083)
Other	0.094	0.052	−0.459
	(−0.331, 0.524)	(−0.393, 0.487)	(−1.386, 0.336)
Years of education	−0.057	−0.130	−0.099
	(−0.103, −0.013)	(−0.176, −0.083)	(−0.178, −0.017)
Log of family income	−0.172	−0.254	−0.348
	(−0.326, −0.024)	(−0.397, −0.116)	(−0.563, −0.119)
Employment (ref. = unemployed)			
Full-time employed	−0.088	−0.671	−0.900
	(−0.758, 0.539)	(−1.230, −0.102)	(−1.901, 0.180)
Part-time employed	0.013	−0.573	−1.195
	(−0.731, 0.715)	(−1.220, 0.067)	(−2.611, 0.136)
Other employment	0.067	−0.220	0.452
	(−0.621, 0.705)	(−0.792, 0.344)	(−0.447, 1.462)
Constant	−3.086	−2.690	−1.312
	(−4.481, −1.690)	(−3.892, −1.407)	(−3.415, 0.928)

Note: Credible intervals are in parentheses, DIC = 3006.232, and N = 1299.

response variable and 10 independent variables, one would have a total of 10 variable-specific omnibus tests and up to 30 pair-wise tests, three for each independent variable. The problem here speaks to the fact that even when all of these tests are independent of each other, simply testing multiple hypotheses simultaneously increases the likelihood of falsely rejecting the null.

Conceptually, Bayesian analysis approaches this problem quite differently by having one joint posterior distribution, which is usually considered a complete, panoramic view of the parameters (Kruschke 2015, pp. 328–329). There are several ways to make Bayesian inferences. We could use a central interval of posterior probabilities, such as the 95% credible intervals used in the previous example. Alternatively, we could use the highest density intervals (HDIs), which are intervals that contain densities higher than those outside the interval (Kruschke 2015). These intervals are often considered more informative than central intervals, especially for examining skewed distributions. At the same time, one could use the region of practical equivalence (ROPE) to help inform Bayesian inference (Spiegelhalter and Freedman 1994; Kruschke 2011). As a few Bayesian statisticians have pointed out (Gelman et al. 2012; Kruschke 2013), hypotheses regarding intervals are more useful than point estimates. For example, the null hypothesis may be that the parameter is between −0.1 and 0.1 rather than equal to 0 (Cohen 1988), and the values that fall within this interval are considered to be "negligibly different from the null for practical purposes" (Kruschke 2013, p. 577).

To illustrate, we will use the self-rated health example from the previous section. Table 6.8 contains results from a Bayesian estimation of a nonparallel cumulative logit model of health. According to the results, although the magnitude of the coefficients varies somewhat across equations, substantive conclusions from these results for most independent variables are similar to those from a parallel cumulative logit model. For example, as one ages, the odds of having poor versus good health increases, and both education and income increase the odds of being in relatively good versus poor health categories.

For comparison, we also estimated parallel and nonparallel cumulative logit models using the classical frequentist approach. The results for the nonparallel cumulative logit model from both approaches are quite similar. This is not surprising given that Bayesian estimation using noninformative priors converges to frequentist estimation. With results from both models, we also tested the parallel regression assumption using the Brant, LR, and Wald tests. The results from separate and joint tests suggest that the coefficients for Black and education violate the parallel regression assumption. An omnibus test also indicates that the assumption is violated for the entire model.

Based on the information from the "HDIs" with the "ROPE" in Figure 6.1, we can safely conclude that the three coefficients for Black have violated the parallel assumption. Specifically, the results indicate that the differences in coefficients for Blacks from the second (3–4 vs. 1–2) and third (4 vs. 1–3) equations and from the first (2–4 vs. 1) and third (4 vs. 1–3) equations are not significantly different from zero. However, this is not the case when comparing the coefficients for Blacks from the first and second equations, and we cannot reach a definitive conclusion because its corresponding 95% HDI does not completely fall within or outside the prespecified ROPE. Here, we set the ROPE to be from −0.1 to 0.1; that is, instead of having one single point for the null value, we chose a range of values to be negligibly different from zero. We also tried other reasonably wider ROPE values, such as −0.5 to 0.5, and reached the same conclusion. Such results actually concur with those from the LR and Wald tests of coefficient equivalence.

Figure 6.2, on the other hand, shows that we have to suspend judgment regarding the parallel assumption for education because none of the 95% interval is completely within or outside the prespecified ROPE, which is set to be from −0.05 to 0.05 given the relatively small magnitude of coefficients. This conclusion diverges from the one provided by the frequentist approach, which indicated that the parallel regression assumption was clearly violated.

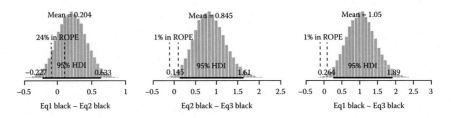

FIGURE 6.1
Highest density intervals with specified region of practical equivalence for Blacks.

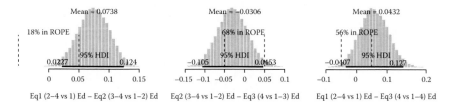

FIGURE 6.2
Highest density intervals with specified region of practical equivalence for education.

This example illustrates that there is a trade-off between having an additional parameter, and therefore possibly more accurate predictions, and the loss of parsimony associated with the addition of each new parameter. If there is strong theory or mounting empirical evidence that suggests a different pattern for a cumulative logit, then relaxing the parallel assumption adds utility to our analyses. Otherwise, there is no real harm in disregarding a modest improvement in model fit, which may well reflect peculiarities in the sample. In adjudicating between the parallel and nonparallel Bayesian models, the popular Bayesian model fit statistic, deviance information criterion (DIC), does provide more support for the nonparallel model (DIC = 3006.232) than the parallel model (DIC = 3021.430).

Recent advances in software applications and computing power have resolved some of the estimation problems in more complex ordered regression models. For example, using the reversible-jump MCMC, McKinley et al. (2015) developed a Bayesian method for fitting the nonparallel cumulative logit model that allows for the "stochastic ordering" constraints or the equivalent of the condition that $\tau_{m-1} - x\beta_{m-1} < \tau_m - x\beta_m$. The method guarantees that the stochastic ordering will hold for the observed range of values (McKinley et al. 2015, p. 10). This is one solution to the potential convergence problems in nonparallel cumulative models with negative predicted probabilities using the traditional, frequentist approach.

In short, there are several advantages associated with the Bayesian approach to statistical inference. First, we do not have to worry about arduous correction or adjustment in multiple comparisons and simultaneous hypothesis testing because there is a single multivariate posterior distribution of parameters. Second, the combined use of HDI and ROPE can largely resolve issues associated with the impractical setup of traditional null hypotheses for point estimates. Third, Bayesian inference can comfortably accept the null, whereas the frequentist approach can only fail to reject the null. Finally, fitting a Bayesian hierarchical model when there are multiple comparisons involved can resolve problems associated with testing the parallel regression assumption in an innovative way (Gelman et al. 2012; Kruschke 2013). These are just a few of the advantages of using a Bayesian approach.

6.6.3 Bayesian Estimation of the Random Cutpoint Multilevel Stereotype Logit Model

Previous sections on the Bayesian estimation of ordered regression models provide examples that the frequentist approach can estimate with similar results using noninformative priors. As we discussed, however, the Bayesian approach can estimate models that might be too complex for traditional methods to reach convergence due to matrix singularities or identification issues. For example, the stereotype logit model is one of the less commonly used ordered response models. It is a constrained form of the well-known multinomial logit model. A stereotype logit model for an ordinal response variable, $y = 1, 2,...,m,...,M$, can be specified as follows

$$\ln\left(\frac{P(y = m \mid \mathbf{x})}{P(y = 1 \mid \mathbf{x})}\right) = \tau_m - \mathbf{x}\phi_m\boldsymbol{\beta}$$

$$\Rightarrow P(y = m \mid \mathbf{x}) = \frac{\exp(\tau_m - \mathbf{x}\phi_m\boldsymbol{\beta})}{\displaystyle\sum_{m=1}^{M} \exp(\tau_m - \mathbf{x}\phi_m\boldsymbol{\beta})} \qquad (6.19)$$

where we must assume that $0 = \phi_1 < \phi_2 <... < \phi_m <... < \phi_M = 1$ in order to retain the ordinality of the relationships (Anderson 1984). In practice, however, the ordering of ϕ's can be hard to implement in estimation using the traditional frequentist approach. Therefore, statistical software packages estimate an unordered stereotype logit model that does not place constraints on the φ parameters.

The Bayesian approach offers a potential solution to this problem. Ahn et al. (2009) propose that instead of estimating ϕ's directly, one could have another set of auxiliary parameters, λ's, which satisfy the following conditions:

$$\lambda_2 = \phi_2, \lambda_m = \phi_m - \phi_{m-1} \text{ for } m = 3,..., M - 1 \qquad (6.20)$$

as such, λ follows a Dirichlet $(1, ..., 1)$ distribution. In this empirical example, we use the 2007 ISSP data with physical activity as the dependent variable and the same set of independent variables that we used in previous sections without Level 2 predictors. To add complexity to this model for illustrative purposes only, we let the cutpoints vary across countries, thereby using τ_{mc} instead of τ_m in (6.19), where m corresponds to a specific threshold and c to a country.

The equation for the joint posterior distribution is

$$p(\tau,\beta,\phi \mid y) \propto p(y \mid \tau,\beta,\phi)p(\tau,\beta,\phi)$$

$$p(y \mid \beta,\phi) = \prod_{i=1}^{N} \frac{\exp(\tau_{mc} - x_i\phi_m\beta)}{\displaystyle\sum_{m=1}^{M}\exp(\tau_{mc} - x_i\phi_m\beta)}$$

$$p(\tau,\beta,\phi) = p(\tau)p(\beta)p(\phi)$$

$$p(\tau_{mc}) \sim N(\bar{\tau}_m + m + 0.5, v_m)\forall\tau_{mc} \qquad (6.21)$$

$$p(\bar{\tau}_m) \sim N(0, 1.0E - 6)\forall\bar{\tau}_m$$

$$p(v_m) \sim \text{Gamma}(1.0E - 3, 1.0E - 3)\forall v_m$$

$$p(\beta) \sim N(0, 1.0E - 6)\forall\beta$$

$$p(\lambda) \sim \text{Dirichlet}(1, 1, 1, 1)$$

Here we specify the priors for λ's due to the parameterization of the model.

Table 6.9 contains results from the Bayesian estimation of a multilevel stereotype logit model of self-reported physical activity. Age decreases the odds of being physically active, whereas education and positive occupational characteristics are associated with an increase in physical activity. In addition, as we change our odds from a low to high level of physical activity in comparison to never ($y = 1$), the magnitude of the effects of independent variables increases as shown by the scale factors going from 0.772 to 0.976. This actually aligns well with our expectation that these independent variables should more strongly differentiate among ordinal response categories that are farther apart.

TABLE 6.9

Bayesian Estimation of a Random Cutpoint Multilevel Stereotype Logit Model of Self-Rated Physical Activity

	$\bar{\beta}$ (95% Credible Interval)
Age (in years)	−0.020
	(−0.029, −0.012)
Sex (male = 1)	0.442
	(0.187, 0.697)
Marital status (married = 1)	−0.113
	(−0.380, 0.154)
Education (ref. = < higher secondary)	
Higher secondary level	0.333
	(0.023, 0.649)
More than higher secondary level	0.664
	(0.332, 1.001)
Employment (ref. = unemployed)	
Other employment type	0.122
	(−0.390, 0.631)
Employed	−0.080
	(−0.577, 0.412)
Locality (city = 1)	−0.040
	(−0.299, 0.219)
Log of family income	0.786 (0.648, 0.935)
Thresholds	
Mean of cutpoint 2 $\bar{\tau}_{02}$	−8.446
	(−9.863, −7.273)
Mean of cutpoint 3 $\bar{\tau}_{03}$	−10.364
	(−11.797, −9.172)
Mean of cutpoint 4 $\bar{\tau}_{04}$	−11.362
	(−12.816, −10.110)
Mean of cutpoint 5 $\bar{\tau}_{05}$	−13.100
	(−14.593, −11.800)
Scale factors	
Factor 2 ϕ_2	0.772
	(0.629, 0.918)
Factor 3 ϕ_3	0.943
	(0.852, 0.995)
Factor 4 ϕ_4	0.976
	(0.920, 0.999)

Note: Credible intervals are in parentheses; $N = 1830$.

6.7 Conclusion

In this chapter, we covered two extensions of traditional ordered regression models: heterogeneous choice models and multilevel ordered regression models. We also briefly introduced Bayesian inference as an alternative to the use of MLE for several ordered regression models. We do not present these model extensions and alternative estimation framework to demonstrate the superiority of more complex models or Bayesian estimation but rather to showcase additional modeling possibilities for ordinal outcomes. Simpler models often fit the data at least as well as relatively complex ordered regression models. Parallel models may suffice in many cases in the social and behavioral sciences if practitioners are more concerned about empirical significance than theoretical assumptions. However, the aforementioned complex models may be preferable over simpler ordered regression models in cases where there is a substantial amount of prediction error, a more complex data structure, or differences in unobserved heterogeneity among key groups. We recommend comparing the results from several ordered models of varying complexity in the model selection process.

References

Agresti, A. 2010. *Analysis of Ordinal Categorical Data*, 2nd ed. Hoboken, NJ: Wiley.

Agresti, A. 2012. *Categorical Data Analysis*, 3rd ed. Hoboken, NJ: Wiley.

Agresti, A. and B. F. Agresti. 1978. Statistical Analysis of Qualitative Variation. *Sociological Methodology* 9: 204–237.

Agresti, A. and D. B. Hitchcock. 2005. Bayesian Inference for Categorical Data Analysis. *Statistical Methods & Applications* 14: 297–330.

Ahn, J., B. Mukherjee, M. Banerjee, and K. A. Cooney. 2009. Bayesian Inference for the Stereotype Regression Model: Application to a Case-Control Study of Prostate Cancer. *Statistics in Medicine* 28: 3139–3157.

Aickin, M. and H. Gensler. 1996. Adjusting for Multiple Testing When Reporting Research Results: The Bonferroni vs Holm Methods. *American Journal of Public Health* 86: 726–728.

Aitchison, J. and S. D. Silvey. 1957. The Generalization of Probit Analysis to the Case of Multiple Responses. *Biometrika* 44: 131–140.

Akaike, H. 1973. Information Theory and an Extension of the Maximum Likelihood Principle. In *Second International Symposium on Information Theory*, edited by B. N. Petrov and F. Csaki. Budapest, Hungary: Akadémiai Kiadó, pp. 267–281.

Albert, A. and J. A. Anderson. 1984. On the Existence of Maximum Likelihood Estimates in Logistic Regression Models. *Biometrika* 71: 1–10.

Albert, J. H. and S. Chib. 2001. Sequential Ordinal Modeling with Applications to Survival Data. *Biometrics* 57: 829–836.

Aldrich, J. H. and F. D. Nelson. 1984. *Linear Probability, Logit, and Probit Models*. Newbury Park, CA: Sage.

Allison, P. D. 1982. Discrete-Time Methods for the Analysis of Event Histories. *Sociological Methodology* 13: 61–98.

Allison, P. D. 1999. Comparing Logit and Probit Coefficients Across Groups. *Sociological Methods and Research* 28: 186–208.

Allison, P. D. 2008. Convergence Failures in Logistic Regression. In *SAS Global Forum 2008*. San Antonio, TX.

Amemiya, T. 1981. Qualitative Response Models: A Survey. *Journal of Economic Literature* 19: 1483–1536.

Anderson, J. A. 1984. Regression and Ordered Categorical Variables. *Journal of the Royal Statistical Society, Series B (Methodological)* 46: 1–30.

Begg, C. B. and R. Gray. 1984. Calculation of Polychotomous Logistic Regression Parameters using Individualized Regressions. *Biometrika* 71: 11–18.

Bera, A. K. and C. R. McKenzie. 1986. Alternative Forms and Properties of the Score Test. *Journal of Applied Statistics* 13: 13–25.

Berkson, J. 1944. Application of the Logistic Function to Bio-Assay. *Journal of the American Statistical Association* 39: 357–365.

Berkson, J. 1951. Why I Prefer Logits to Probits. *Biometrics* 7: 327–339.

Blair, J. and M. G. Lacy. 2000. Statistics of Ordinal Variation. *Sociological Methods and Research* 28: 251–280.

Bliss, C. I. 1934a. The Method of Probits. *Science* 79: 38–39.

Bliss, C. I. 1934b. The Method of Probits: A Correction. *Science* 79: 409–410.

Boes, S. and R. Winkelmann. 2010. The Effect of Income on General Life Satisfaction and Dissatisfaction. *Social Indicators Research* 95: 111–128.

Brant, R. 1990. Assessing Proportionality in the Proportional Odds Model for Ordinal Logistic Regression. *Biometrics* 46: 1171–1178.

Breen, R., A. Holm, and K. B. Karlson. 2014. Correlations and Nonlinear Probability Models. *Social Methods & Research* 43: 571–605.

Breen, R., R. Luijkx, W. Müller, and R. Pollak. 2009. Nonpersistent Inequality in Educational Attainment: Evidence from Eight European Countries. *American Journal of Sociology* 114: 1475–1521.

Breusch, T. S. and A. R. Pagan. 1980. The Lagrange Multiplier Test and Its Applications to Model Specification in Econometrics. *Review of Economic Studies* 47: 239–253.

Buis, M. L. 2007. SEQLOGIT: Stata Module to Fit a Sequential Logit Model. http://ideas.repec.org/c/boc/bocode/s456843.html. Accessed on 9/20/14.

Buis, M. L. 2011. The Consequences of Unobserved Heterogeneity in a Sequential Logit Model. *Research in Social Stratification and Mobility* 29: 247–262.

Buis, M. L. 2013. Oparallel: Stata Module Providing Post-Estimation Command for Testing the Parallel Regression Assumption. http://econpapers.repec.org/software/bocbocode/s457720.html. Accessed on 7/26/15.

Buis, M. L. 2015. Not All Transitions Are Equal: The Relationship Between Effects on Passing Steps in a Sequential Process and Effects on the Final Outcome. *Sociological Methods and Research* (In Press).

Burnham, K. P. and D. R. Anderson. 2004. Multimodel Inference: Understanding AIC and BIC in Model Selection. *Sociological Methods and Research* 33: 261–304.

Buse, A. 1982. The Likelihood Ratio, Wald, and Lagrange Multiplier Tests: An Expository Note. *American Statistician* 36: 153–157.

Cacioppo, J. T. and G. G. Berntson. 1994. Relationship between Attitudes and Evaluative Space: A Critical Review, With Emphasis on the Separability of Positive and Negative Substrates. *Psychological Bulletin* 115: 401–423.

Cacioppo, J. T., W. L. Gardner, and G. G. Berntson. 1997. Beyond Bipolar Conceptualizations and Measures: The Case of Attitudes and Evaluative Space. *Personality and Social Psychology Review* 1: 3–25.

Cameron, S. V. and J. J. Heckman. 1998. Life Cycle Schooling and Dynamic Selection Bias: Models and Evidence for Five Cohorts of American Males. *Journal of Political Economy* 106: 262–333.

Capuano, A. W. and J. D. Dawson. 2013. The Trend Odds Model for Ordinal Data. *Statistics in Medicine* 32: 2250–2261.

Carlin, B. P. and T. A. Louis. 2009. *Bayesian Methods for Data Analysis*. New York: CRC Press.

Cheng, S. and J. S. Long. 2007. Testing for IIA in the Multinomial Logit Model. *Sociological Methods and Research* 35: 583–600.

Choi, T. and S. R. Cole. 2004. A Family of Ordered Logistic Regression Models Fit by Data Expansion. *International Journal of Epidemiology* 33: 1413.

Clogg, C. C. and E. S. Shihadeh. 1994. *Statistical Models for Ordinal Variables*. Thousand Oaks, CA: Sage.

Cohen, J. 1988. *Statistical Power Analysis for the Behavioral Sciences*. Hillsdale, NJ: Erlbaum.

Cole, S. R. and C. V. Ananth. 2001. Regression Models for Unconstrained, Partially or Fully Constrained Continuation Odds Ratios. *International Journal of Epidemiology* 30: 1379–1382.

Cole, S. R., P. D. Allison, and C. V. Ananth. 2004. Estimation of Cumulative Odds Ratios. *Annals of Epidemiology* 14: 172–178.

Eliason, S. R. 1993. *Maximum Likelihood Estimation: Logic and Practice*. Newbury Park, CA: Sage.

Engle, R. F. 1984. Wald, Likelihood Ratio, and Lagrange Multiplier Tests in Econometrics. In *Handbook of Econometrics II* (pp. 776–826), edited by Z. Griliches and M. D. Intriligator. New York: North-Holland.

Fienberg, S. E. 1980. *The Analysis of Cross-Classified Categorical Data*, 2nd ed. Cambridge, MA: MIT Press.

Finney, D. J. 1947. *Probit Analysis*. Cambridge, UK: Cambridge University Press.

Finney, D. J. 1952. *Probit Analysis*, 2nd ed. Cambridge, UK: Cambridge University Press.

Finney, D. J. 1971. *Probit Analysis*, 3rd ed. Cambridge, UK: Cambridge University Press.

Fisher, R. A. 1935. Appendix to Bliss, C. I.: The Case of Zero Survivors. *Annals of Applied Biology* 22: 164–165.

Fullerton, A. S. 2009. A Conceptual Framework for Ordered Logistic Regression Models. *Sociological Methods and Research* 38: 306–347.

Fullerton, A. S. and K. F. Anderson. 2013. The Role of Job Insecurity in Explanations of Racial Health Inequalities. *Sociological Forum* 28: 308–325.

Fullerton, A. S. and J. C. Dixon. 2009. Racialization, Asymmetry, and the Context of Welfare Attitudes in the American States. *Journal of Political and Military Sociology* 37: 95–120.

Fullerton, A. S. and J. C. Dixon. 2010. Generational Conflict or Methodological Artifact? Reconsidering the Relationship between Age and Policy Attitudes in the U.S., 1984–2008. *Public Opinion Quarterly* 74: 643–673.

Fullerton, A. S., J. C. Dixon, and C. Borch. 2007. Bringing Registration into Models of Vote Overreporting. *Public Opinion Quarterly* 71: 649–660.

Fullerton, A. S. and M. J. Stern. 2013. Racial Differences in the Gender Gap in Political Participation in the American South, 1952–2004. *Social Science History* 37: 145–176.

Fullerton, A. S. and J. Xu. 2012. The Proportional Odds with Partial Proportionality Constraints Model for Ordinal Response Variables. *Social Science Research* 41: 182–198.

Fullerton, A. S. and J. Xu. 2015. Constrained and Unconstrained Partial Adjacent Category Logit Models for Ordinal Response Variables. *Sociological Methods & Research* (In Press).

Garwood, F. 1941. The Application of Maximum Likelihood to Dosage-Mortality Curves. *Biometrika* 32: 46–58.

Gauchat, G. 2011. The Cultural Authority of Science: Public Trust and Acceptance of Organized Science. *Public Understanding of Science* 20: 751–770.

Gelman, A., J. B. Carlin, H. S. Stern, D. B. Dunson, A. Vehtari, and D. B. Rubin. 2014. *Bayesian Data Analysis*, 3rd ed. Boca Raton, FL: CRC Press.

Gelman, A. and J. Hill. 2007. *Data Analysis Using Regression and Multilevel/Hierarchical Models*. Cambridge, UK: Cambridge University Press.

Gelman, A., J. Hill, and M. Yajima. 2012. Why We (Usually) Don't Have to Worry About Multiple Comparisons. *Journal of Research on Educational Effectiveness* 5: 189–211.

Gelman, A. and C. P. Robert. 2013a. "Not Only Defended but Also Applied": The Perceived Absurdity of Bayesian Inference. *The American Statistician* 67: 1–5.

Gelman, A. and C. P. Robert. 2013b. Rejoinder: The Anti-Bayesian Moment and Its Passing. *The American Statistician* 67: 16–17.

Gilens, M. 1999. *Why Americans Hate Welfare: Race, Media, and the Politics of Antipoverty Policy*. Chicago, IL: University of Chicago Press.

Gill, J. 2015. *Bayesian Methods: A Social and Behavioral Sciences Approach*, 3rd ed. Boca Raton, FL: Chapman & Hall/CRC Press.

Goodman, L. A. 1983. The Analysis of Dependence in Cross-Classifications Having Ordered Categories, Using Log-Linear Models for Frequencies and Log-Linear Models for Odds. *Biometrics* 39: 149–160.

Greene, W. H. and D. A. Hensher. 2010. *Modeling Ordered Choices: A Primer*. Cambridge, UK: Cambridge University Press.

Greenland, S. 1994. Alternative Models for Ordinal Logistic Regression. *Statistics in Medicine* 13: 1665–1677.

Haberman, S. J. 1982. Analysis of Dispersion of Multinomial Responses. *Journal of the American Statistical Association* 77: 568–580.

Halaby, C. N. 1986. Worker Attachment and Workplace Authority. *American Sociological Review* 51: 634–649.

Hanmer, M. J. and K. O. Kalkan. 2013. Behind the Curve: Clarifying the Best Approach to Calculating Predicted Probabilities and Marginal Effects from Limited Dependent Variable Models. *American Journal of Political Science* 57: 263–277.

Hastie, T., R. Tibshirani, and J. Friedman. 2009. *The Elements of Statistical Learning: Data Mining, Inference, and Prediction*, 2nd ed. New York, NY: Springer.

Hauser, R. M. and M. Andrew. 2006. Another Look at the Stratification of Educational Transitions: The Logistic Response Model with Partial Proportionality Constraints. *Sociological Methodology* 36: 1–26.

Heckman, J. J. 1979. Sample Selection Bias as a Specification Error. *Econometrica* 47: 153–161.

Hedeker, D. and R. Gibbons. 1994. A Random-Effects Ordinal Regression Model for Multilevel Analysis. *Biometrics* 50: 933–944.

Hedeker, D., R. J. Mermelstein, and K. A. Weeks. 1999. The Thresholds of Change Model: An Application to Analyzing Stages of Change Data. *Annals of Behavioral Medicine* 21: 61–70.

Holm, A. and M. M. Jæger. 2011. Dealing with Selection Bias in Educational Transition Models: The Bivariate Probit Selection Model. *Research in Social Stratification and Mobility* 29: 311–322.

House, J. S., J. M. Lepkowski, A. M. Kinney, R. P. Mero, R. C. Kessler, and A. R. Herzog. 1994. The Social Stratification of Aging and Health. *Journal of Health and Social Behavior* 35: 213–234.

ISSP Research Group. 2009. *International Social Survey Programme: Leisure Time and Sports—ISSP 2007.* edited by G. D. Archive. Cologne, Germany: ISSP Research Group.

Jones, B. S. and M. E. Sobel. 2000. Modeling Direction and Intensity in Semantically Balanced Ordinal Scales: An Assessment of Congressional Incumbent Approval. *American Journal of Political Science* 44: 174–185.

Jordan, N. 1965. The 'Asymmetry' of 'Liking' and 'Disliking': A Phenomenon Meriting Further Reflection and Research. *Public Opinion Quarterly* 29: 315–322.

Kahn, L. M. and K. Morimune. 1979. Unions and Employment Stability: A Sequential Logit Approach. *International Economic Review* 20: 217–235.

Kawachi, I., S. V. Subramanian, and D. Kim. 2010. *Social Capital and Health.* New York: Springer.

Keele, L. and D. K. Park. 2006. *Difficulty Choices: An Evaluation of Heterogeneous Choice Models.* Oxford, UK: Nuffield College, Oxford University.

Kendzor, D. E., L. R. Reitzel, C. A. Mazas, L. M. Cofta-Woerpel, Y. Cao, L. Ji, T. J. Costello, et al. 2012. Individual- and Area-Level Unemployment Influence Smoking Cessation among African Americans Participating in a Randomized Clinical Trial. *Social Science & Medicine* 74: 1394–1401.

Kim, J. 2003. Assessing Practical Significance of the Proportional Odds Assumption. *Statistics & Probability Letters* 65: 233–239.

Kruschke, J. K. 2011. Bayesian Assessment of Null Values via Parameter Estimation and Model Comparison. *Perspectives on Psychological Science* 6: 299–312.

Kruschke, J. K. 2013. Bayesian Estimation Supersedes the T Test. *Journal of Experimental Psychology: General* 142: 573–603.

Kruschke, J. K. 2015. *Doing Bayesian Data Analysis: A Tutorial with R, JAGS, and Stan,* 2nd ed. San Diego, CA: Elsevier.

Laara, E. and J. N. S. Matthews. 1985. The Equivalence of Two Models for Ordinal Data. *Biometrika* 72: 206–207.

Lacy, M. G. 2006. An Explained Variation Measure for Ordinal Response Models with Comparisons to Other Ordinal R^2 Measures. *Sociological Methods and Research* 34: 469–520.

Lall, R., M. J. Campbell, S. J. Walters, K. Morgan, and MRC CFAS Co-operative. 2002. A Review of Ordinal Regression Models Applied on Health-Related Quality of Life Assessments. *Statistical Methods in Medical Research* 11: 49–67.

Laplace, P. S. 1986. Memoir on the Probability of the Causes of Events. *Statistical Science* 1: 364–378.

Liao, T. F. 1994. *Interpreting Probability Models: Logit, Probit, and Other Generalized Linear Models.* Thousand Oaks, CA: Sage.

Liao, T. F. 2004. Comparing Social Groups: Wald Statistics for Testing Equality among Multiple Logit Models. *International Journal of Comparative Sociology* 45: 3–16.

Liao, T. F. and G. Stevens. 1994. Spouses, Homogamy, and Social Networks. *Social Forces* 73: 693–707.

Lieberson, S. 1985. *Making It Count: The Improvement of Social Research and Theory.* Berkeley, CA: University of California Press.

Long, J. S. 1997. *Regression Models for Categorical and Limited Dependent Variables.* Thousand Oaks, CA: Sage.

Long, J. S. 2009. *Group Comparisons in Logit and Probit Using Predicted Probabilities.* Bloomington, IN: Indiana University.

Long, J. S. 2015. Regression Models for Nominal and Ordinal Outcomes. In *The Sage Handbook of Regression Analysis and Causal Inference*, edited by H. Best and C. Wolf. Thousand Oaks, CA: Sage, pp. 173–203.

Long, J. S. and S. Cheng. 2004. Regression Models for Categorical Outcomes. In *Handbook of Data Analysis*, edited by M. Hardy and A. Bryman. Thousand Oaks, CA: Sage, pp. 259–284.

Long, J. S. and J. Freese. 2014. *Regression Models for Categorical Dependent Variables Using Stata*, 3rd ed. College Station, TX: Stata Press.

Lynch, S. M. 2010. *Introduction to Applied Bayesian Statistics and Estimation for Social Scientists*. New York, NY: Springer.

Maddala, G. S. 1983. *Limited-Dependent and Qualitative Variables in Econometrics*. Cambridge, UK: Cambridge University Press.

Mare, R. D. 1980. Social Background and School Continuation Decisions. *Journal of the American Statistical Association* 75: 295–305.

Mare, R. D. 1981. Change and Stability in Educational Stratification. *American Sociological Review* 46: 72–87.

Mare, R. D. 2006. Response: Statistical Models of Educational Stratification—Hauser and Andrew's Models for School Transitions. *Sociological Methodology* 36: 27–37.

Mare, R. D. 2011. Introduction to Symposium on Unmeasured Heterogeneity in School Transition Models. *Research in Social Stratification and Mobility* 29: 239–245.

McCullagh, P. 1980. Regression Models for Ordinal Data. *Journal of the Royal Statistical Society Series B* 42: 109–142.

McCullagh, P. and J. A. Nelder. 1989. *Generalized Linear Models*, 2nd ed. New York, NY: Chapman and Hall.

McFadden, D. 1973. Conditional Logit Analysis of Qualitative Choice Behavior. In *Frontiers in Econometrics*, edited by P. Zarembka. New York, NY: Academic Press, pp. 105–142.

McKelvey, R. D. and W. Zavoina. 1975. A Statistical Model for the Analysis of Ordinal Level Dependent Variables. *Journal of Mathematical Sociology* 4: 103–120.

McKinley, T. J., M. Morters, and J. L. N. Wood. 2015. Bayesian Model Choice in Cumulative Link Ordinal Regression Models. *Bayesian Analysis* 10: 1–30.

Mood, C. 2010. Logistic Regression: Why We Cannot Do What We Think We Can Do, and What We Can Do About It. *European Sociological Review* 26: 67–82.

O'Connell, A. A. 2006. *Logistic Regression Models for Ordinal Response Variables*. Thousand Oaks, CA: Sage.

Olsen, R. J. 1982. Independence from Irrelevant Alternatives and Attrition Bias: Their Relation to One Another in the Evaluation of Experimental Programs. *Southern Economic Journal* 49: 521–535.

Peterson, B. and F. E. Harrell, Jr. 1990. Partial Proportional Odds Models for Ordinal Response Variables. *Applied Statistics* 39: 205–217.

Pfarr, C., A. Schmid, and U. Schneider. 2011. Estimating Ordered Categorical Variables Using Panel Data: A Generalised Ordered Probit Model with an Autofit Procedure. *Journal of Economics and Econometrics* 54: 7–23.

Plummer, M. 2003. JAGS: A Program for Analysis of Bayesian Graphical Models Using Gibbs Sampling. In *Proceedings of the 3rd International Workshop on Distributed Statistical Computing*, pp. 20–22.

Powers, D. A. and Y. Xie. 2008. *Statistical Methods for Categorical Data Analysis*, 2nd ed. Howard House, UK: Emerald.

Pratt, J. W. 1981. Concavity of the Log Likelihood. *Journal of the American Statistical Association* 76: 103–106.

Rabe-Hesketh, S. and A. Skrondal. 2012. *Multilevel and Longitudinal Modeling Using Stata*, 3rd ed. College Station, TX: Stata Press.

Raftery, A. E. 1995. Bayesian Model Selection in Social Research. *Sociological Methodology* 25: 111–163.

Rao, C. R. 1948. Large Sample Tests of Statistical Hypotheses Concerning Several Parameters with Applications to Problems of Estimation. *Mathematical Proceedings of the Cambridge Philosophical Society* 44: 50–57.

Raudenbush, S. W. and A. S. Bryk. 2002. *Hierarchical Linear Models: Application and Data Analysis Methods*, 2nd ed. Thousand Oaks, CA: Sage.

R Core Development Team. 2014. *R: A Language and Environment for Statistical Computing*. R Foundation for Statistical Computing, Vienna, Austria.

Silvey, S. D. 1959. The Lagrange multiplier Test. *Annals of Mathematical Statistics* 30: 389–407.

Smith, T. W., M. Hout, and P. V. Marsden. 2013. *General Social Survey, 1972–2012* [Cumulative File]. edited by Roper Center for Public Opinion Research. Ann Arbor, MI: Inter-University Consortium for Political and Social Research.

Snell, E. J. 1964. A Scaling Procedure for Ordered Categorical Data. *Biometrics* 20: 592–607.

Sobel, M. E. 1997. Modeling Symmetry, Asymmetry, and Change in Ordered Scales with Midpoints Using Adjacent Category Logit Models for Discrete Data. *Sociological Methods and Research* 26: 213–232.

Sobel, M. E., M. P. Becker, and S. M. Minick. 1998. Origins, Destinations, and Association in Occupational Mobility. *American Journal of Sociology* 104: 687–721.

Spiegelhalter, D. J. and L. S. Freedman. 1994. Bayesian Approaches to Randomized Trials. *Journal of Royal Statistical Association. Series A (Statistics in Society)* 157: 357–416.

Sturgis, P., C. Roberts, and P. Smith. 2014. Middle Alternatives Revisited: How the Neither/Nor Response Acts as a Way of Saying 'I Don't Know'? *Sociological Methods and Research* 43: 15–38.

Terza, J. V. 1985. Ordered Probit: A Generalization. *Communications in Statistics—A. Theory and Methods* 14: 1–11.

Train, K. E. 2009. *Discrete Choice Methods with Simulation*, 2nd ed. Cambridge, UK: Cambridge University Press.

Tsai, C. 2013. *Score Tests for Likelihood-Based Models Using Stata*. Department of Government, University of Essex, Colchester, UK. Unpublished Manuscript.

Tutz, G. 1991. Sequential Models in Categorical Regression. *Computational Statistics & Data Analysis* 11: 275–295.

Tutz, G. 2012. *Regression for Categorical Data*. Cambridge, UK: Cambridge University Press.

Vuong, Q. H. 1989. Likelihood Ratio Tests for Model Selection and Non-Nested Hypotheses. *Econometrica* 57: 307–333.

Wald, A. 1943. Tests of Statistical Hypotheses Concerning Several Parameters When the Number of Observations Is Large. *Transactions of the American Mathematical Society* 54: 426–482.

Walker, S. H. and D. B. Duncan. 1967. Estimation of the Probability of an Event as a Function of Several Independent Variables. *Biometrika* 54: 167–179.

Weakliem, D. L. 2004. Introduction to the Special Issue on Model Selection. *Sociological Methods and Research* 33: 167–187.

Wilks, S. S. 1938. The Large-Sample Distribution of the Likelihood Ratio for Testing Composite Hypotheses. *Annals of Mathematical Statistics* 9: 60–62.

Williams, R. 2006. Generalized Ordered Logit/Partial Proportional Odds Models for Ordinal Dependent Variables. *Stata Journal* 6: 58–82.

Williams, R. 2009. Using Heterogeneous Choice Models to Compare Logit and Probit Coefficients across Groups. *Sociological Methods & Research* 37: 531–559.

Williams, R. 2010. Fitting Heterogeneous Choice Models with OGLM. *Stata Journal* 10: 540–567.

Williams, R. 2012. Using the Margins Command to Estimate and Interpret Adjusted Predictions and Marginal Effects. *Stata Journal* 12: 308–331.

Williams, R. 2015. Understanding and Interpreting Generalized Ordered Logit Models. *Journal of Mathematical Sociology* (In Press).

Winship, C. and R. D. Mare. 1984. Regression Models with Ordinal Variables. *American Sociological Review* 49: 512–525.

Wolfe, R. 1998. sg86: Continuation-Ratio Models for Ordinal Response Data. *Stata Technical Bulletin* 44: 18–21.

Wooldridge, J. M. 2010. *Econometric Analysis of Cross Section and Panel Data*, 2nd ed. Cambridge, MA: The MIT Press.

Xie, Y. 2011. Values and Limitations of Statistical Models. *Research in Social Stratification and Mobility* 29: 343–349.

Xu, J. and A. S. Fullerton. 2014. Comparing, Confounding, or Clarifying? Alternative Measures of Statistical Group Comparisons in Binary Regression Models. *Chinese Sociological Review* 46: 91–119.

Yee, T. W. 2010. The VGAM Package for Categorical Data Analysis. *Journal of Statistical Software* 32: 1–34.

Yee, T. W. 2015. *Vector Generalized Linear and Additive Models: With an Implementation in R*. New York, NY: Springer.

Yee, T. W. and T. J. Hastie. 2003. Reduced-Rank Vector Generalized Linear Models. *Statistical Modelling* 3: 15–41.

Index